**MODULAR SERIES
ON SOLID STATE DEVICES**

VOLUME I

Semiconductor Fundamentals

MODULAR SERIES
ON SOLID STATE DEVICES
Robert F. Pierret and Gerold W. Neudeck, Editors

VOLUME I

Semiconductor Fundamentals

ROBERT F. PIERRET

Purdue University

ADDISON-WESLEY PUBLISHING COMPANY

READING, MASSACHUSETTS
MENLO PARK, CALIFORNIA
LONDON • AMSTERDAM
DON MILLS, ONTARIO
SYDNEY

This book is in the
Addison-Wesley Modular Series on Solid State Devices

Library of Congress Cataloging in Publication Data

Pierret, Robert F.
  Semiconductor fundamentals.

  (Modular series on solid state devices ; v. 1)
  Bibliography: p.
  Includes index.
  1. Semiconductors.   I. Title.   II. Series: Pierret,
Robert F. Modular series on solid state devices ; v. 1.
TK 7871.85.P485   621.3815'2   81-14978
ISBN 0-201-05320-9   AACR2

Reprinted with corrections, July 1983

ISBN 0-201-05320-9
GHIJ-AL-898765

Foreword

Solid state devices have attained a level of sophistication and economic importance far beyond the highest expectations of their inventors. The bipolar and field-effect transistors have virtually made possible the computer industry which in turn has created a completely new consumer market. By continually offering better performing devices at lower cost per unit, the electronics industry has penetrated markets never before addressed. One fundamental reason for such phenomenal growth is the enhanced understanding of basic solid state device physics by the modern electronics designer. Future trends in electronic systems indicate that the digital system, circuit, and IC layout design functions are being merged into one. To meet such present and future needs we have written this series of books aimed at a qualitative and quantitative understanding of the most important solid state devices.

Volumes I through IV are written for a junior, senior, or possibly first-year graduate student who has had a reasonably good background in electric field theory. With some deletions these volumes have been used in a one semester, three credit-hour, junior-senior level course in electrical engineering at Purdue University. Following this course are two integrated-circuit-design and two IC laboratory courses. Each volume is written to be covered in 12 to 15 fifty-minute lectures.

The individual volumes make the series useful for adoption in standard and non-standard format courses, such as minicourses, television, short courses, and adult continuing education. Each volume is relatively independent of the others, with certain necessary formulas repeated and referenced between volumes. This flexibility enables one to use the series continuously or in selected parts, either as a complete course or as an introduction to other subjects. We also hope that students, practicing engineers, and scientists will find the volumes useful for individual instruction, whether it is for reference, review, or home study.

A number of the standard texts on devices have been written like encyclopedias, packed with information, with little thought to how the student learns or reasons. Texts that are encyclopedic in nature are difficult for students to read and are often barriers to their understanding. By breaking the material into smaller units of information and by writing for students (rather than for our colleagues) we hope to enhance their understanding of solid state devices. A secondary pedagogical strategy is to strike a healthy balance between the device physics and practical device information.

The problems at the end of each chapter are important to understanding the concepts presented. Many problems are extensions of the theory or are designed to reinforce particularly important topics. Some numerical problems are included to give the reader an intuitive feel for the size of typical parameters. Then, when approximations are stated or assumed, the student will have confidence that certain quantities are indeed orders of magnitude smaller than others. The problems have a range of difficulty, from very simple to quite challenging. We have also included discussion questions so that the reader is forced into qualitative as well as quantitative analyses of device physics.

Problems, along with answers, at the end of the first three volumes represent typical test questions and are meant to be used as review and self-testing. Many of these are discussion, sketch, or "explain why" types of questions where the student is expected to relate concepts and synthesize ideas.

We feel that these volumes present the basic device physics necessary for understanding many of the important solid state devices in present use. In addition, the basic device concepts will assist the reader in learning about the many exotic structures presently in research laboratories that will likely become commonplace in the future.

West Lafayette, Indiana   R. F. Pierret
                                                                         G. W. Neudeck

Contents

3 Carrier Action

Introduction

The primary purpose of this volume is to concisely present and examine those terms, concepts, equations, models, etc., that are routinely employed in describing the operational behavior of solid state devices. In essence, the reader will be exposed to the basics of the language spoken by the solid state device specialist, a language that must be mastered if the reader and the device specialist are to communicate efficiently. The first chapter in the volume provides a general physical description of semiconductors, surveying, for example, their elemental makeup, atomic arrangements, and similar topics. The second chapter presents a detailed description of the current carrying entities inside a semiconductor under "rest" (equilibrium) conditions. The third and final chapter examines what happens to the current carrying entities when the semiconductor is electrically perturbed, and how the resulting situation is described mathematically in terms of observables and known parameters.

The reader should be forewarned that the subject matter covered in this volume is moderately challenging and at times somewhat conceptual in nature. However, in this work every effort has been expended to make the presentation as lucid as possible without jeopardizing the integrity of the topic. One final comment: It should be noted that the sum total of the subject matter has been reduced to the absolute minimum required for a basic understanding of devices.

1 / Semiconductors — General Information

1.1 GENERAL MATERIAL PROPERTIES

The vast majority of all solid state devices on the market today are fabricated from a class of materials known as semiconductors. It therefore seems appropriate that we begin the discussion by examining the general nature of semiconducting materials.

1.1.1 Composition

Most of the semiconductors commonly encountered are listed in Table 1.1. The semiconducting group of materials includes elemental semiconductors such as Si, compound semiconductors such as GaAs, and alloys like $GaAs_xP_{1-x}$.* Of the many cited semiconductors, Si is by far the most important, and at the present time it totally dominates the commercial market. All integrated circuits (ICs) and most devices are made with this material. Needless to say, we will focus our attention on Si in subsequent discussions. GaAs and $GaAs_xP_{1-x}$ are perhaps the most important of the remaining materials. GaAs, exhibiting superior materials properties, may someday challenge Si in the commercial marketplace. $GaAs_xP_{1-x}$ is worthy of note because of its use in visible light-emitting diodes (LEDs). The remaining semiconductors are primarily utilized in highly specialized applications.

Although the number of semiconducting materials is reasonably large, the list is actually quite abbreviated considering the number of elements and possible combinations of elements. As it turns out, referring to Table 1.2, we see that only a certain group of elements and elemental combinations give rise to semiconducting materials. Specifically, all of the semiconductors listed in Table 1.1 are elements appearing in Column IV of the Periodic Table, or they are a combination of elements in Periodic Table columns an equal distance to either side of Column IV. The Column III element

---

*The x in alloy formulas is a fraction lying between 0 and 1, where for example, $GaAs_{0.6}P_{0.4}$ would indicate a material with 6 As and 4 P atoms per every 10 Ga atoms.

Table 1.1. Semiconductor Materials.

General Classification	Specific Examples
(1) Elemental Semiconductors . .	Si, Ge
(2a) III-V Compounds	GaAs, GaP, GaSb, InAs, InP, InSb, AlAs, AlP, AlSb
(2b) II-VI Compounds	CdS, CdSe, CdTe, HgS, ZnO, ZnS, ZnSe, ZnTe
(3) Alloys	$GaAs_xP_{1-x}$, $HgCd_xTe_{1-x}$

Ga plus the Column V element As yields the III–V semiconductor GaAs; the Column II element Zn plus the Column VI element O yields the II–VI semiconductor ZnO. This very general property is, of course, related to the chemical bonding in semiconductors, where, on the average, there are four valence electrons per atom.

1.1.2 Purity

As will be explained in Chapter 2, extremely minute traces of impurity atoms can have a drastic effect on the electrical properties of semiconductors. For this reason, the compositional purity of semiconductors must be very carefully controlled and, in fact, modern semiconductors are some of the purest solid materials in existence. In Si, for example, the unintentional content of non-Si atoms, other than electrically inactive

Table 1.2. Relationship between Semiconductor Composition and the Periodic Table of Elements/ Summarizing Comments.

Column in the Periodic Table of Elements					Comments
II	III	IV	V	VI	
.	.	.			
.	.	Ⓢⓘ – – – – – – –			Si, the most important semiconducting material. ICs and most devices are made of this material — an *elemental* semiconductor.
.	.				
.	Ⓖⓐ + + + + Ⓐⓢ = = =				GaAs, the most widely employed *compound* semiconductor — used in infrared LEDs, Gunn diodes, etc.
.	.	+			
.	.	+			
.	.	+			
.		Ⓟ = = =			$GaAs_xP_{1-x}$, presently the most important *alloy* semiconductor — used in visible LEDs
.					
Ⓩⓝ + + + + + + + + + Ⓞ =					ZnO, used in ceramic varistors and SAW devices.

carbon and oxygen, are held to less than one atom per 10^9 Si atoms. To assist the reader in attempting to comprehend this incredible level of purity, let us suppose a forest of maple trees was planted from coast to coast, border to border, at 50-ft centers across the United States (including Alaska). An impurity level of one part per 10^9 corresponds to finding about 25 crabapple trees in the maple tree forest covering the United States! It should be emphasized that the cited material purity refers to *unintentional* undesired impurities. Typically, specially selected impurity atoms at levels of one part per 10^8 to one impurity atom per 10^3 semiconductor atoms will be *purposely* added to the semiconductor to control its electrical properties.

1.1.3 Structure

The spatial arrangement of atoms within any material plays an important role in determining the precise properties of the material. As shown schematically in Fig. 1.1, the atomic arrangement within solids falls into one of three broad classifications; namely, amorphous, polycrystalline, and crystalline. An amorphous solid is a material in which there is no recognizable long-range order in the positioning of atoms within the material. The atomic arrangement in any given section of an amorphous material will look different from the atomic arrangement in any other section of the material. Crystalline solids lie at the opposite end of the "order" spectrum; in a crystalline material the atoms are arranged in an orderly three-dimensional array. Given any section of a crystalline material, one can readily reproduce the atomic arrangement in any other section of the material. Polycrystalline solids comprise an intermediate case in which the solid is composed of crystalline subsections that are disjointed or misoriented relative to each other.

Examining the gamut of solid state devices in production today, one can find applications involving each of the three cited structural forms. The overwhelming majority

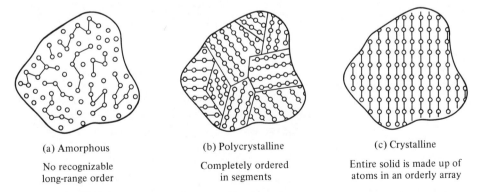

(a) Amorphous

No recognizable
long-range order

(b) Polycrystalline

Completely ordered
in segments

(c) Crystalline

Entire solid is made up of
atoms in an orderly array

Fig. 1.1 General classification of solids based on the degree of atomic order: (a) amorphous, (b) polycrystalline, and (c) crystalline.

of semiconductor materials in common usage, however, are *crystalline*—the over-whelming number of devices fabricated today employ *crystalline* semiconductors.

1.2 CRYSTAL LATTICES

The discussion at the end of the preceding section leads nicely into the topic of this section. Since in-use semiconductors are typically crystalline in form, it seems reasonable to seek out additional information about the crystalline state. Our major goal here is to present a more detailed picture of the atomic arrangement within the principal semiconductors. To achieve this goal, we first examine how one goes about describing the spatial positioning of atoms within crystals. Next, a bit of "visualization" practice with simple three-dimensional lattices (atomic arrangements) is presented prior to examining semiconductor lattices themselves. The section concludes with an introduction to the related topic of Miller indices. Miller indices are a sort of shorthand notation widely employed for identifying specific planes and directions within crystals.

1.2.1 The Unit Cell Concept

Simply stated, a unit cell is a small portion of any given crystal that could be used to reproduce the crystal. To help establish the unit cell (or building-block) concept, let us consider the two-dimensional lattice shown in Fig. 1.2(a). In order to describe this lattice or to totally specify the physical characteristics of the lattice, one need only provide the unit cell shown in Fig. 1.2(b). As indicated in Fig. 1.2(c), the original lattice can be readily reproduced by merely duplicating the unit cell and stacking the duplicates next to each other in an orderly fashion.

In dealing with unit cells there often arises a misunderstanding, and hence confusion, relative to two points. First of all, unit cells are not necessarily unique. The unit cell shown in Fig. 1.2(d) is as acceptable as the Fig. 1.2(b) unit cell for specifying the original lattice of Fig. 1.2(a). Second, a unit cell need not be primitive (the smallest unit cell possible). In fact, it is usually advantageous to employ a slightly larger unit cell with orthogonal sides instead of a primitive cell with nonorthogonal sides. This is especially true in three dimensions where noncubic unit cells are quite difficult to describe and visualize.

1.2.2 Simple 3-D Unit Cells

The real world is three-dimensional and so are semiconductor crystals. Naturally, the unit cells required to describe semiconductor crystals are also three-dimensional. In Fig. 1.3(a) we have pictured the simplest of all three-dimensional unit cells; namely, the simple cubic unit cell. The simple cubic cell, you will note, is an equal-sided box or cube with an atom positioned at each corner of the cube. The simple cubic lattice associated with this cell is best visualized by referring to Fig. 1.2(a). Repeating the two-dimensional lattice of Fig. 1.2(a) at spacings of "a" out of the paper and positioning the atoms in successive planes directly over those in the preceding plane yields

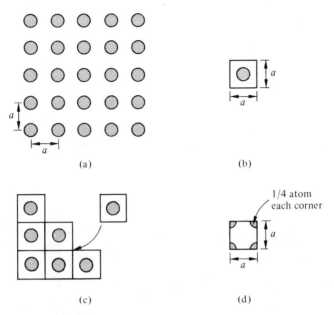

Fig. 1.2 Introducing the unit cell method of describing atomic arrangements within crystals. (a) Sample two-dimensional lattice. (b) Unit cell corresponding to the part (a) lattice. (c) Reproducing the original lattice. (d) An alternative unit cell.

the simple cubic lattice. Alternatively, one can of course build up the simple cubic lattice using the simple cubic unit cell. In doing so, however, it should be noted that only ⅛ of each corner atom is actually *inside* the unit cell, as pictured in Fig. 1.3(b). If the Fig. 1.3(b) unit cell is now duplicated and stacked like play blocks in a nursery, one again derives the simple cubic lattice.

Figures 1.3(c) and 1.3(d) display two common 3-D unit cells that are somewhat more complex but still closely related to the simple cubic cell. In the Fig. 1.3(c) unit cell an atom is added at the center of the cube; this configuration is appropriately called the body centered cubic (bcc) unit cell. The face centered cubic (fcc) unit cell of Fig. 1.3(d) contains an atom at each face of the cube in addition to the atoms at each corner. (Note, however, that only one-half of each face atom actually lies inside the fcc unit cell.) Whereas the simple cubic cell contains one atom (⅛ of an atom at each of the eight cube corners), the somewhat more complex bcc and fcc cells contain two and four atoms, respectively. It is left to the reader to verify these facts and to visualize the lattices associated with bcc and fcc cells.

1.2.3 The Diamond Lattice

We are finally in a position to provide additional details relative to the positioning of atoms within the principal semiconductors. In Si (and Ge) the lattice structure is de-

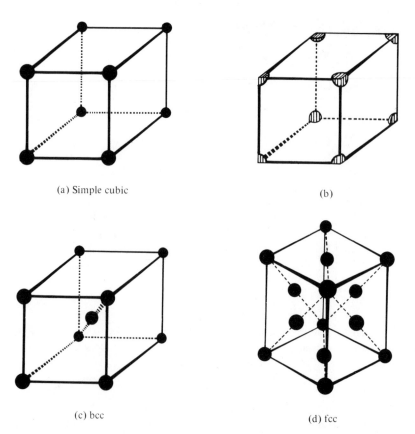

(a) Simple cubic

(b)

(c) bcc

(d) fcc

Fig. 1.3   Simple three-dimensional unit cells. (a) Simple cubic unit cell. (b) Pendantically correct simple cubic unit cell including only the fractional portion (⅛) of each corner atom actually within the cell cube. (c) Body centered cubic unit cell. (d) Face centered cubic unit cell. [(a), (c), and (d) from *Physics of Solids* by C. A. Wert and R. M. Thomson, ©1964 by McGraw-Hill Book Co., New York.]

scribed by the unit cell pictured in Fig. 1.4(a). The Fig. 1.4(a) arrangement is known as the *diamond lattice* unit cell because it also characterizes diamond, a form of the Column IV element carbon. Examining the diamond lattice unit cell, we note that the cell is cubic, with atoms at each corner and at each face of the cube similar to the fcc cell. The interior of the Fig. 1.4(a) cell, however, contains four additional atoms. One of the interior atoms is located along a cube body diagonal exactly one-quarter of the way down the diagonal from the top front left-hand corner of the cube. The other three interior atoms are displaced one-quarter of the body diagonal length along the previously noted body diagonal direction from the front, top, and left-side face atoms, respectively. The really astute reader will realize we have just described the diamond lattice as nothing more than two interpenetrating fcc lattices, where one fcc lattice is

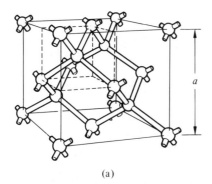

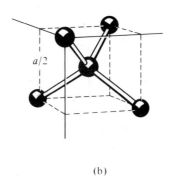

(a)   (b)

Fig. 1.4 (a) The diamond lattice unit cell. For Si the cube side length is 5.43Å. (b) Enlarged top corner of the part (a) diamond lattice, emphasizing the four-nearest-neighbor bonding within the structure. (From *Electrons and Holes in Semiconductors* by William Shockley, ©1950 by Litton Educational Publishing, Inc. Reprinted by permission of Van Nostrand Reinhold Co.)

displaced one-quarter of the body diagonal length along a body diagonal direction relative to the other fcc lattice. To complete the picture, it should be noted that GaAs (and the other III – V semiconductor compounds) crystallize in the so-called zinc-blende lattice, which is essentially identical to the diamond lattice except that each of the component elements is restricted to positions on only one of the above cited inter-penetrating fcc sublattices. In other words, Ga would take the place of the face and corner atoms in Fig. 1.4(a) while As would be substituted for the four interior atoms.

Now that the positioning of atoms within the principal semiconductors has been established, the question may arise as to the practical utilization of such information. Although several applications could be cited, geometrical-type calculations constitute a very common and readily explained use of the unit cell formalism. For example, in Si at room temperature the unit cell side length (a) is 5.43 Å (where 1 Å = 10^{-8} cm). Since there are eight Si atoms per unit cell and the volume of the unit cell is a^3, it follows that there are $8/a^3$ or almost exactly 5×10^{22} atoms/cm^3 in the Si lattice. Similar calculations could be performed to determine atomic radii, the distance between atomic planes, etc. For the purposes of the development herein, however, the major reason for the unit cell discussion was to establish that, as emphasized in Fig. 1.4(b), *atoms in the diamond lattice have four nearest neighbors*. The chemical bonding (or crystalline glue) within the major semiconductors is therefore dominated by the attrac-tion between any given atom and its four closest neighbors. This is a rather important fact that should be filed away for future reference.

1.2.4 Miller Indices

Single crystals of silicon used in device processing normally assume the form and shape pictured in Fig. 1.5. This single crystal of silicon, better known as a Si wafer, is commonly ~15 mil thick (1 mil = 0.001 inch) and 2 to 5 in. in diameter, with at least one surface carefully polished and etched to yield a damage-free, mirrorlike finish. Of particular interest here is the fact that the wafer's surface is also carefully preoriented to lie along a specific crystallographic plane, and a "flat" is ground along the periphery of the wafer to identify a reference direction within the surface plane. Precise surface orientation is critical in certain device processing steps and directly affects the characteristics exhibited by some devices; the "flat" is routinely employed to orient device arrays on the wafer so as to achieve high yields during device separation. The point we wish to make is that specification of crystallographic planes and directions is of practical importance. Miller indices, exemplified by the (100) plane and [011] direction designations used in Fig. 1.5, constitute the conventional means of identifying planes and directions within a crystal.

The Miller indices for any given plane of atoms within a crystal are obtained by following a straightforward four-step procedure. The procedure is detailed below, along with the simultaneous sample indexing of the plane shown in Fig. 1.6(a).

Indexing Procedure	Sample Implementation
After setting up coordinate axes along the edges of the unit cell, note where the plane to be indexed intercepts the axes. Divide each intercept value by the unit cell length along the respective coordinate axis. Record the resulting normalized (pure-number) intercept set in the order x, y, z.	2, 1, 3
Invert the intercept values; that is, form [1/intercept]s.	½, 1, ⅓
Using an appropriate multiplier, convert the 1/intercept set to the smallest possible set of whole numbers.	3, 6, 2
Enclose the whole-number set in curvilinear brackets.	(362)

To complete the description of the indexing procedure, the user should also be aware of the following special facts:

i) If the plane to be indexed is parallel to a coordinate axis, the "intercept" along the axis is taken to be at infinity. Thus, for example, the plane shown in Fig. 1.6(b) intercepts the coordinate axes at 1, ∞, ∞, and is therefore a (100) plane.

ii) If the plane to be indexed has an intercept along the negative portion of a coordinate axis, a ($^{-}$) is placed *over* the corresponding index number. In the Miller indexing scheme, then, the Fig. 1.6(c) plane is designated a ($1\bar{1}1$) plane.

iii) Looking back to the diamond lattice of Fig. 1.4(a), note that the six planes passing through the cube faces contain identical atom arrangements; that is, because of crystal symmetry, it is impossible to distinguish between the "equivalent" (100), (010), (001), ($\bar{1}$00), ($0\bar{1}0$), and ($00\bar{1}$) planes; or, it is impossible to distinguish between {100} planes. A group of equivalent planes is concisely referenced in the Miller notation through the use of { } braces.

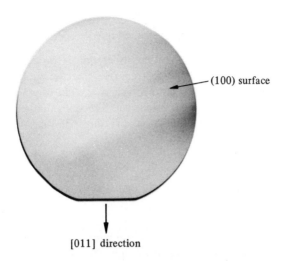

[011] direction

Fig. 1.5 Single crystal silicon wafer with a (100) surface and a [011] flat. Si wafers with (100) and (111) surfaces are standard; other orientations can be obtained only by special order. The figure is primarily intended to dramatize the utility of Miller indices.

iv) Miller indices enclosed within square brackets, [ ], are used to designate directions within a crystal; triangular brackets, ⟨ ⟩, designate equivalent directions. For the cubic class of crystals (which encompasses the major semiconductors), an [hkl] direction is normal to the corresponding (hkl) plane. Hence, without an extensive amount of explanation, it should be clear that the x-axis is a [100] direction, [1̄10]

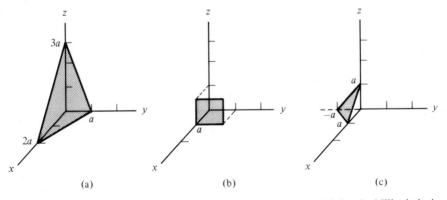

Fig. 1.6 Sample cubic crystal planes. (a) The (362) plane used in explaining the Miller indexing procedure. (b) The (100) plane. (c) The (1̄1̄1) plane.

is a direction in the xy-plane displaced $45°$ from the x-axis toward the negative y-axis, $[111]$ is a direction along the cube body diagonal, etc.

Having concluded an examination of the Miller indices formalism, we would like to observe that the formalism is not all that complex; becoming comfortable with the indexing scheme, however, does require a certain amount of practice. In addition to obtaining the Miller indices for given planes and directions, the reader should practice the reverse process—picturing the planes and directions described by given sets of indices.

1.3 CRYSTAL GROWTH

1.3.1 Obtaining Ultrapure Si

Considering the obvious general availability and widespread use of semiconductor materials, particularly Si, it is reasonable to ask how the ultrapure, single-crystal Si used in modern-day device production is obtained in the first place. Could it be that Si is readily available in sandstone deposits?—No. Perhaps, as a sort of by-product, Si single crystals come from South African diamond mines?—Wrong again. As suggested recently in a low-budget science fiction movie, maybe Si is scooped from the ocean bottom by special submarines?—Sorry, no. Although Si is the second most abundant element in the earth's crust and a component in numerous compounds, chief of which are silica (impure SiO_2) and the silicates (Si + O + another element), silicon never occurs alone in nature as an element. Single crystal Si used in device production, it turns out, is a specially made material.

Given the above introduction, it should be clear that the initial steps in producing device-quality silicon must involve separating Si from its compounds and purifying the separated material. The ingenious separation and purification process that has evolved is summarized in Fig. 1.7. Low-grade silicon or ferrosilicon is first produced by heating silica with carbon in an electric furnace. The carbon essentially pulls the oxygen away from the impure SiO_2 (i.e., *reduces* the SiO_2), leaving behind impure Si. Next, the ferrosilicon is chlorinated to yield $SiCl_4$ or $SiHCl_3$, both of which are liquids at room temperature. Although appearing odd at first glance, the liquefaction process is actually a brilliant maneuver. Whereas solids are very difficult to purify, a number of standard procedures are available for purifying liquids. An ultrapure $SiCl_4$ (or $SiHCl_3$) is the result after multiple distillation and other liquid purification procedures. Lastly, the high purity halide is chemically reduced, yielding the desired ultrapure elemental silicon. This is accomplished, for example, by heating $SiCl_4$ in a hydrogen atmosphere $[SiCl_4 + 2H_2 \rightarrow 4HCl + Si]$.

1.3.2 Single Crystal Formation

Although ultrapure, the silicon derived from the separation and purification process as just described is not a single crystal, but is instead polycrystalline. Additional

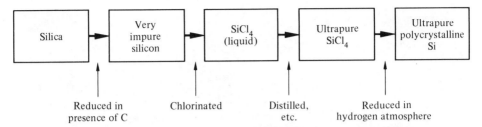

Fig. 1.7   Summary of the process employed to produce ultrapure silicon.

processing is therefore required to form the large single crystals used in device fabrication. The most commonly employed method yielding large single crystals of Si is known as the Czochralski pulled method. In this method the ultrapure polycrystalline silicon is placed in a quartz crucible and heated in an inert atmosphere to form a melt, as shown in Fig. 1.8. A small single crystal, or Si "seed" crystal, with the normal to its bottom face carefully aligned along a predetermined direction (typically a $\langle 111 \rangle$ or $\langle 100 \rangle$

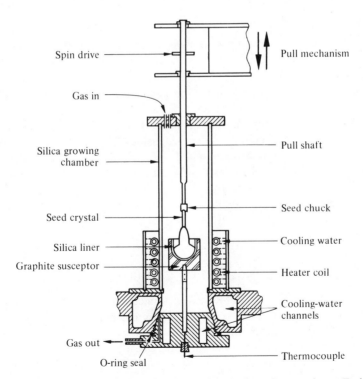

Fig. 1.8   Typical silicon crystal-pulling apparatus. (From *Silicon Semiconductor Technology* by W. R. Runyan, ©1965 by McGraw-Hill Book Co., New York.)

Fig. 1.9   Silicon ingots and crystal growth materials. Pictured from left to right on the table are: Back row — three quartz crucibles in varying sizes, graphite crucible holder, and crucible filled with polycrystalline Si chunks, bag of polycrystalline Si, portion of a 4″ Si ingot. Second through fourth rows — 3″, 1½″, and 2″ Si ingots, respectively. Fifth row — 4″ Si ingot, portion of a 4″ Si crystal, saw blade used in slicing the ingots into wafers. Front row — package of dopant material and a seed crystal. (Photograph courtesy of Delco Electronics Div., GMC, Kokomo, Indiana.)

direction), is then clamped to a metal rod and dipped into the melt. Once thermal equilibrium is established, the temperature of the melt in the vicinity of the seed crystal is reduced, and silicon from the melt begins to freeze out onto the seed crystal, the added material being a structurally perfect extension of the seed crystal. Subsequently, the seed crystal is slowly rotated and withdrawn from the melt; this permits more and more silicon to freeze out on the bottom of the growing crystal. The resulting large, cylindrically shaped single crystal of silicon, also known as an ingot (see Fig. 1.9), is typically 2 to 5 in. in diameter and up to 35 in. in length. The Si wafers used in device processing (Fig. 1.5) are ultimately produced by cutting the ingot into thin sections using a diamond-edged saw.

1.4   SUMMARY AND CONCLUDING COMMENTS

In this chapter the reader was introduced to semiconductors in general and silicon in particular. As is the usual case in the initial examination of a subject, most of our attention was directed to gross features. We hope, however, that the reader now understands that semiconductors are ultrapure and typically highly ordered materials, has gained some proficiency in visualizing and describing crystals, is familiar with the atomic bonding scheme inside the diamond lattice, and is knowledgeable about the production of large single crystals of silicon.

PROBLEMS

1.1 Answer the following questions as concisely as possible.

(a) What is the difference between a crystalline and a polycrystalline material?

(b) How many atoms are there in a simple cubic unit cell? — in a bcc unit cell? — in a fcc unit cell? — in the unit cell characterizing the diamond lattice?

(c) What is being indicated by the bracket sets ( ), [ ], { }, and ⟨ ⟩ as employed in the Miller indexing scheme?

(d) Describe the Czochralski pulled method for obtaining large single crystals of silicon.

1.2 If the lattice constant or unit cell side length in Si is $a = 5.43 \times 10^{-8}$cm, what is the distance from the center of one Si atom to the center of its nearest neighbor?

1.3 Germanium crystallizes in the diamond lattice and contains 4.42×10^{22} atoms/cm^3. Determine the unit cell side length for this semiconductor.

1.4 (a) Consider a {100} plane, a plane containing the face of a Si unit cell. What is the number of atoms per cm^2 on this plane; i.e., how many atoms per cm^2 have their centers on a {100} plane?

(b) Repeat for the (110) plane.

1.5 Assuming a cubic crystal system, make a sketch of the following planes.

(a) (010)   (b) (111)   (c) (321)   (d) ($1\bar{1}0$)

1.6 Assuming a cubic crystal system, use an appropriately directed arrow to identify each of the following directions.

(a) [001]   (b) [110]   (c) [$0\bar{1}0$]   (d) [111]

1.7 In a cubic lattice, how many equivalent directions are associated with each of the following designations?

(a) ⟨100⟩   (b) ⟨110⟩   (c) ⟨111⟩

1.8 Treating atoms as rigid spheres with radii equal to one-half the distance between nearest neighbors, show that the ratio of the volume occupied by the atoms to the total available volume in the various crystal structures is:

(a) $\pi/6$ or 52% for the simple cubic lattice,

(b) $\sqrt{3}\,\pi/8$ or 68% for the body-centered cubic lattice,

(c) $\sqrt{2}\,\pi/6$ or 74% for the face-centered cubic lattice,

(d) $\sqrt{3}\,\pi/16$ or 34% for the diamond lattice.

2 / Carrier Modeling

Carriers are the entities that transport charge from place to place inside a material and hence give rise to electrical currents. In everyday life the most commonly encountered type of carrier is the electron, the subatomic particle responsible for charge transport in metallic wires. Within semiconductors one again encounters the familiar electron, but there is also a second equally important type of carrier—the hole. Electrons and holes are the focal point of this chapter, wherein we examine carrier related concepts, models, properties, and terminology.

Although reminders will be periodically interjected, it should be emphasized from the very start that the development throughout this entire chapter assumes that *equilibrium* conditions exist within the semiconductor. Equilibrium, in layman's terms (and we really can't do much better at this point), refers to a semiconductor that is just sitting there and has been sitting there for a long time. There are no external voltages, electric fields, magnetic fields, stresses, or what have you applied to the semiconductor; and all internal observables are invariant with time. This "rest" condition, as it turns out, provides an excellent *frame of reference*. Knowing what the system is like when it is "just sitting there" permits one to extrapolate and ascertain the system's condition when a perturbation has been applied.

As a final introductory comment, at certain points in this chapter we will pull formulas or facts right "out of the air" and will insist that the reader accept the formulas or facts without justification. We would like to give a more complete discussion, properly developing every concept and deriving every formula, but unfortunately this is not possible. (Indeed, entire books have been written on what we have omitted.) Moreover, it is our underlying philosophy that being able to derive a result is secondary to knowing how to *use* a result. The motivated reader can, of course, fill in any information gap through supplemental reading.

2.1 THE QUANTIZATION CONCEPT

Instead of attempting to deal immediately with electrons in crystalline Si, where there are 14 electrons per atom and 5×10^{22} atoms/cm^3, we will take a more realistic approach and first establish certain ground rules by examining much simpler atomic systems. We begin with the simplest of all atomic systems, the isolated hydrogen atom. This atom, as the reader may recall from a course in modern physics, was under intense scrutiny at the start of the 20th century. Scientists of that time knew the hydrogen atom contained a negatively charged particle in orbit about a more massive positively charged nucleus. What they could not explain was the nature of the light emitted from the system when the hydrogen atom was heated to an elevated temperature. Specifically, the emitted light was observed at only certain discrete wavelengths; according to the prevailing theory of the time, scientists expected a continuum of wavelengths.

In 1913 Niels Bohr proposed a solution to the dilemma. Bohr hypothesized that the H-atom electron was restricted to certain well-defined orbits; or, equivalently, Bohr assumed that the orbiting electron could take on only certain values of angular momentum. This "quantization" of the electron's angular momentum was, in turn, coupled directly to energy quantization. As can be readily established, if the electron's angular momentum is assumed to be $\mathbf{n}\hbar$, then

$$E_H = -\frac{m_0 q^4}{2(4\pi\varepsilon_0\hbar\mathbf{n})^2} = -\frac{13.6}{\mathbf{n}^2}\ \text{eV}, \qquad \mathbf{n} = 1, 2, 3, \ldots \qquad (2.1)$$

where (also see Fig. 2.1) E_H is the electron binding energy within the hydrogen atom, m_0 is the mass of a free electron, q is the magnitude of the electronic charge, ε_0 is the permittivity of free space, h is Planck's constant, $\hbar = h/2\pi$, and $\mathbf{n}$ is the energy quantum number or orbit identifier. The *electron volt* (eV) is a unit of energy equal to 1.6×10^{-19} joules. Now, with the electron limited to certain energies inside the hydrogen atom, it follows from the Bohr model that the transition from a higher $\mathbf{n}$ to a lower $\mathbf{n}$ orbit will release quantized energies of light; this explains the observation of emitted light at only certain discrete wavelengths.

For our purposes, the most important idea to be obtained from the Bohr model is that the energy of electrons in atomic systems is restricted to a limited set of values.* Relative to the hydrogen atom, the energy level scheme in a multi-electron atom like silicon is, as one might intuitively expect, decidedly more complex. It is still a relatively easy task, however, to describe the salient energy-related features of an isolated silicon atom. As pictured in Fig. 2.2, ten of the 14 Si-atom electrons occupy very deep-lying energy levels and are tightly bound to the nucleus of the atom. The binding is so strong, in fact, that these ten electrons remain essentially unperturbed during chemical reactions or normal atom–atom interactions, with the ten-electron-plus-nucleus combination often being referred to as the "core" of the atom. The remaining four Si-atom electrons, on the other hand, are rather weakly bound and are collectively called the *valence electrons* because of their strong participation in chemical reactions. As emphasized in Fig. 2.2, the four valence electrons, if unperturbed, occupy four of the eight allowed slots (or states) having

---

*Actually, it is now well established that not only energy but many other observables relating to atomic-sized particles are quantized, and an entire field of study, quantum mechanics, has been developed to describe the properties and actions of atomic-sized particles and systems.

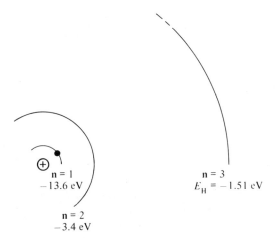

Fig. 2.1   The hydrogen atom — idealized representation showing the first three allowed electron orbits and the associated energy quantization.

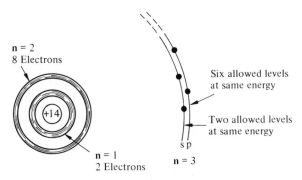

Fig. 2.2   Schematic representation of an isolated Si atom.

the next highest energy above the deep-lying core levels. Finally, we should note, for completeness, that the electronic configuration in the 32-electron Ge-atom (germanium being the other elemental semiconductor) is essentially identical to the Si-atom configuration except the Ge-core contains 28 electrons.

2.2 SEMICONDUCTOR MODELS

Building on the information presented in previous sections, we introduce and describe in this section two very important models or visualization aids which are used extensively in the analysis of semiconductor devices. The inclusion of semiconductor models in a chapter devoted to carrier modeling may appear odd at first glance but is actually quite appropriate. We are, in effect, modeling the carrier "container," the semiconductor crystal.

2.2.1 Bonding Model

The isolated Si atom, or a Si atom not interacting with other atoms, was found to contain four valence electrons. Si atoms incorporated in the diamond lattice, on the other hand, exhibit a bonding that involves an attraction between each atom and its four nearest neighbors (refer to Fig. 1.4(b)). The implication here is that in going from isolated atoms to the collective crystalline state Si atoms come to share one of their valence electrons with each of the four nearest neighbors. This covalent bonding, or equal sharing of valence electrons with nearest neighbors, and the mere fact that atoms in the diamond lattice have four nearest neighbors, give rise to the idealized semiconductor representation, the bonding model, shown in Fig. 2.3. Each circle in the Fig. 2.3 bonding model represents the core of a semiconductor atom, while each line represents a shared valence electron. (There are eight lines connected to each atom because any given atom not only *contributes* four shared electrons but must also *accept* four shared electrons from adjacent atoms.) The two-dimensional nature of the model is, of course, an idealization that facilitates mental visualizations and also makes it much easier to reproduce the model on paper and chalkboards.

Although considerable use will be made of the bonding model in subsequent discussions, it is nevertheless worthwhile at this point to examine sample applications of the model to provide some indication of the model's utility. Two sample applications are presented in Fig. 2.4. In Fig. 2.4(a) we use the bonding model to picture a point defect, a missing atom, in the lattice structure. In Fig. 2.4(b) we visualize the breaking of an atom-to-atom bond and the associated release or freeing of an electron. [Bond breaking (at $T > 0°K$) and defects occur naturally in all semiconductors, and hence (if we may be somewhat overcritical) the basic model of Fig. 2.3 is strictly valid for an entire semiconductor only at $T \simeq 0°K$ when the semiconductor in question contains no defects and no impurity atoms.]

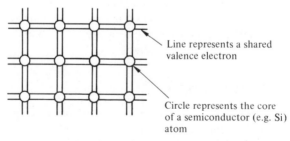

Line represents a shared valence electron

Circle represents the core of a semiconductor (e.g. Si) atom

Fig. 2.3   The bonding model.

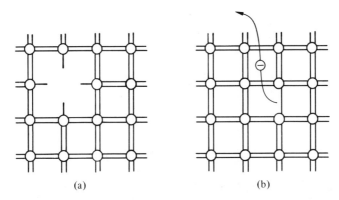

Fig. 2.4 Sample utilization of the bonding model. (a) Visualization of a missing atom or point defect. (b) Breaking of an atom-to-atom bond and freeing of an electron.

2.2.2 Energy Band Model

If our interests were restricted to describing the spatial aspects of events taking place within a semiconductor, the bonding model alone would probably be adequate. Quite often, however, one is more interested in the energy-related aspects of an event. In such instances the bonding model, which says nothing about electron energies, is of little value and the energy band model becomes the primary visualization aid.

Let us begin the conceptual path leading to the energy band model by recalling the situation inside an isolated Si atom. Reviewing the Section 2.1 discussion, ten of the 14 electrons inside an isolated Si atom are tightly bound to the nucleus and are unlikely to be significantly perturbed by normal atom–atom interactions. The remaining four electrons are rather weakly bound and, if unperturbed, occupy four of the eight allowed energy states immediately above the last core level. Moreover, it is implicitly understood that the electronic energy states within a group of Si atoms, say N Si atoms, are all identical — as long as the atoms are isolated, that is, far enough apart so that they are noninteracting.

Given the foregoing knowledge of the isolated atom situation, the question next arises as to whether we can use the knowledge to deduce information about the crystalline state. Assuredly we can omit any further mention of the core electrons because these electrons are not significantly perturbed by normal interatomic forces. The opposite, however, is true of the valence electrons. If N atoms are brought into close proximity (the case in crystalline Si), it is quite reasonable to expect a modification in the energy states of the valence electrons.

The modification in the valence-electron energy states actually known to take place is summarized in Fig. 2.5. In going from N isolated Si atoms to an N-atom Si crystal, exactly half of the allowed states become depressed in energy and half increase in energy. The perturbation, moreover, causes a spread in allowed energies, forming two ranges,

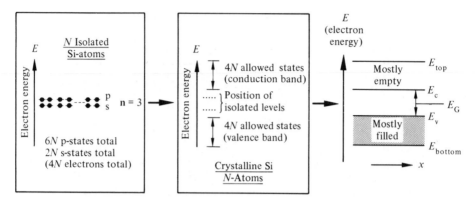

Fig. 2.5 Conceptual development of the energy band model starting with N isolated Si atoms on the left and concluding with a "dressed-up" version of the energy band model on the right.

or *bands*, of allowed energy states separated by an intervening energy gap. The upper band of allowed states is called the *conduction band*; the lower band of allowed states, the *valence band*; and the intervening energy gap, the *forbidden gap* or *band gap*. In filling the allowed energy band states, electrons, of course, tend to gravitate to the lowest possible energies. Noting that electrons are restricted to single occupancy in allowed states (the Pauli exclusion principle) and remembering that the $4N$ valence band states can just accommodate what were formerly $4N$ valence electrons, we typically find that the valence band is almost completely filled with electrons and the conduction band is all but devoid of electrons. Indeed, the valence band is completely filled and the conduction band completely empty at temperatures approaching $T = 0°K$.

To complete our plausibilization of the energy band model, we need to introduce and utilize one additional fact; namely, unlike the valence electrons in the isolated-atom case, the band electrons in crystalline silicon are not tied to or associated with any one particular atom. True, on the average one will typically find four electrons being shared between any given Si atom and its four nearest neighbors (as in the bonding model). However, the identity of the shared electrons changes as a function of time, with the electrons moving around from point to point in the crystal. In other words, the allowed electronic states are no longer atomic states but are associated with the crystal as a whole; independent of the point examined in a perfect crystal, one sees the same allowed-state configuration. We therefore conclude that for a perfect crystal under equilibrium conditions a plot of the allowed electron energies versus distance along any preselected crystalline direction (always called the x-direction) is as shown on the right-hand side of Fig. 2.5. The cited plot, a plot of allowed electron energy states as a function of position, is the basic energy band model. E_c introduced in the Fig. 2.5 plot is the lowest possible conduction band energy, E_v is the highest possible valence band energy, and $E_G = E_c - E_v$ is the band gap energy.

Finally, Fig. 2.6 displays the form of the energy band model (for a perfect crystal under equilibrium conditions) actually employed in practice. In this widely employed "shorthand" version, the line to indicate the top energy in the conduction band, the line to indicate the bottom energy in the valence band, the cross-hatching to indicate filled states, the labeling of the y- or electron-energy axis, and the labeling of the x- or position axis are all understood to exist implicitly, but are not shown explicitly.

2.2.3 Carriers

With the semiconductor models having been firmly established, we are at long last in a position to introduce and to visualize the current carrying entities within semiconductors. Referring to Fig. 2.7, we note first of all from part (a) that, if there are no broken bonds in the bonding model, or equivalently, if in the energy band model the valence band is completely filled with electrons and the conduction band is devoid of electrons, then there are no carriers. Valence band electrons in the energy band model correspond to shared electrons in the bonding model and these electrons are not involved in charge transport.

The electrons involved in charge transport are visualized in Fig. 2.7(b). When a Si – Si bond is broken and the associated electron is free to wander about the lattice, the released electron is a carrier. Equivalently, in terms of the energy band model, excitation of valence band electrons into the conduction band creates carriers; that is, *electrons in the conduction band are carriers*. Note that the energy required to break a bond in the bonding model and the band gap energy, E_G, are one and the same thing. Likewise, freed bonding-model electrons and conduction band electrons are just different names for the same electrons; in subsequent discussions the word "electrons," when used without a modifier, will be understood to refer to these conduction band electrons.

In addition to releasing an electron, the breaking of a Si – Si bond also creates a missing bond or void in the bonding structure. Thinking in terms of the bonding model, one can visualize the movement of this missing bond from place to place in the lattice as a result of nearby bound electrons jumping into the void (see Fig. 2.7(c)). Alternatively, one can think in terms of the energy band model where the removal of an electron from the valence band creates an empty state in an otherwise vast sea of filled states. The empty state, like a bubble in a liquid, moves about rather freely in the lattice because of the cooperative motion of the valence band electrons. What we have been describing, the missing bond

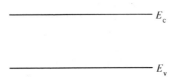

Fig. 2.6 The energy band diagram—widely employed simplified version of the energy band model.

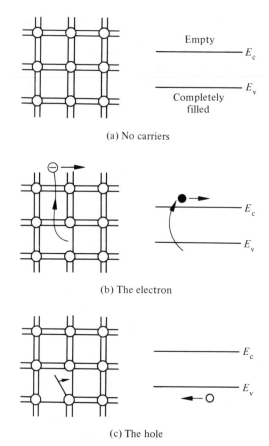

(a) No carriers

(b) The electron

(c) The hole

Fig. 2.7 Visualization of carriers using the bonding model (left) and the energy band model (right). (a) No-carrier situation; (b) visualization of an electron; (c) visualization of a hole.

in the bonding scheme, the empty state in the valence band, is the second type of carrier found in semiconductors — the hole.

Although the preceding description would appear to belie the fact, the valence band hole is actually an entity on an equal footing with the conduction band electron. As one's familiarity with carrier modeling grows, the comparable status of electrons and holes becomes more and more apparent. Eventually, the reader will come to think of the hole as just another subatomic particle. How can a hole be both a particle and a void? The conceptual dilemma here is a direct result of the almost unavoidable inadequacy of our models. The bonding model in particular is a gross oversimplification of a complex situation and exhibits obvious flaws when examined closely. Truthfully, it is a wonder our models work as well as they do; we are attempting to characterize in everyday terms a system that is beyond the realm of everyday life. Yes, in "reality" the hole is just another subatomic "particle."

2.2.4 Material Classification

Changing the subject somewhat, we would like to conclude this section by citing a very important tie between the band gap of a material, the number of carriers available for transport in a material, and the overall nature of a material. As it turns out, although specifically established for semiconductors, the energy band model of Fig. 2.6 can be applied with only slight modification to all materials. The major difference between materials lies not in the nature of the energy bands, but rather in the magnitude of the energy band gap.

Insulators, as illustrated in Fig. 2.8(a), are characterized by very wide band gaps, with E_G for the insulating materials diamond and SiO_2 being $\simeq 5$ eV and $\simeq 8$ eV, respectively. In these wide band gap materials the thermal energy available at room temperature excites very few electrons from the valence band into the conduction band; thus very few carriers exist inside the material and the material is therefore a poor conductor of current. The band gap in metals, by way of comparison, is either very small, or no band gap exists at all due to an overlap of the valence and conduction bands (see Fig. 2.8(c)). There is always an abundance of carriers in metals, and hence metals are excellent conductors. Semiconductors present an intermediate case between insulators and metals. At room

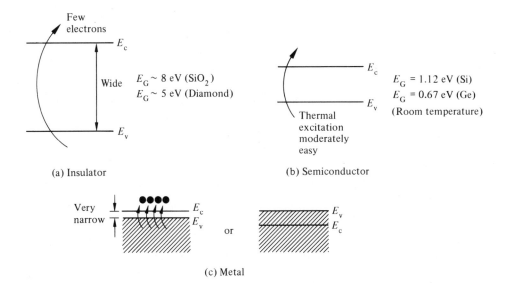

Fig. 2.8 Explanation of the distinction between (a) insulators, (b) semiconductors, and (c) metals using the energy band model.

temperature, $E_G = 1.12$ eV in Si and $E_G = 0.67$ eV in Ge. Thermal energy, by exciting electrons from the valence band into the conduction band, creates a moderate number of carriers in these materials, giving rise in turn to a current-carrying capability intermediate between poor and excellent.

From what we have said, the band gap of a material is clearly an important parameter. Actually, it is perhaps *the* most important material parameter; and the value of the silicon band gap at room temperature ($E_G = 1.12$ eV) should definitely be committed to memory.

2.3   CARRIER PROPERTIES

Having formally introduced the electron and the hole, we next seek to learn as much as possible about the nature of these carriers. In this particular section we examine a collage of carrier-related information, information of a general nature including general facts, properties, and terminology.

2.3.1   Charge

Both electrons and holes are charged entities. Electrons are negatively charged, holes are positively charged, and the *magnitude* of the carrier charge, q, is the same for the two types of carriers. In MKS units, $q = 1.6 \times 10^{-19}$ coul. (Please note that, under the convention adopted herein, the electron and hole charges are $-q$ and $+q$, respectively; i.e., the sign of the charge is displayed explicitly.)

2.3.2   Effective Mass

Mass, like charge, is another very basic property of electrons and holes. Unlike charge, however, the carrier mass is not a simple property and cannot be disposed of by simply quoting a number. Indeed, the apparent or effective mass of electrons within a crystal is a function of the semiconductor material (Si, Ge, etc.) and is different from the mass of electrons within a vacuum.

Seeking to obtain insight into the effective mass concept, let us first consider the motion of electrons in a vacuum. If, as illustrated in Fig. 2.9(a), an electron of rest mass m_0 is moving in a vacuum between two parallel plates under the influence of an electric field $\mathscr{E}$, then, according to Newton's second law, the force $\mathbf{F}$ on the electron will be

$$\mathbf{F} = -q\mathscr{E} = m_0 \frac{d\mathbf{v}}{dt} \qquad (2.2)$$

where $\mathbf{v}$ is the electron velocity and t is time. Next consider electrons (conduction band electrons) moving between the two parallel end faces of a semiconductor crystal under the influence of an applied electric field, as envisioned in Fig. 2.9(b). Does Eq. (2.2) also describe the overall motion of electrons within the semiconductor crystal? The answer is clearly *no*. Electrons moving inside a semiconductor crystal will collide with semiconductor atoms, thereby causing a periodic deceleration of the carriers. However, should not Eq. (2.2) apply to the portion of the electronic motion occurring *between* collisions?

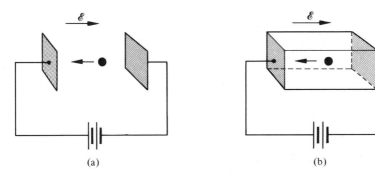

Fig. 2.9 An electron moving in response to an applied electric field (a) within a vacuum, and (b) within a semiconductor crystal.

The answer is again *no*. In addition to the applied electric field, electrons in a crystal are also subject to complex crystalline fields not specifically included in Eq. (2.2).

The foregoing discussion delineated certain important differences between electrons in a crystal and electrons in a vacuum, but left unresolved the equally important question as to how one properly describes the motion of carriers in a crystal. Strictly speaking, the motion of carriers in a crystal can only be described using quantum mechanics, the formalism appropriate for atomic-sized systems. Fortunately, however, if one examines the carrier motion occurring between collisions, the mathematically complex quantum-mechanical formulation simplifies to yield an equation of motion identical to Eq. (2.2), except m_0 is replaced by an effective carrier mass. In other words, for the Fig. 2.9(b) electrons one can write

$$\mathbf{F} = -q\mathscr{E} = m_n^* \frac{d\mathbf{v}}{dt}, \qquad (m_n^* \rightarrow m_p^* \quad \text{for holes}) \qquad (2.3)$$

where m_n^* is the electron effective mass. The internal crystalline fields and quantum-mechanical effects are all suitably lumped into the effective mass factor multiplying $d\mathbf{v}/dt$. The significance of this result is that

> *both conceptually and mathematically, electrons and holes can be treated as classical particles, as long as the carrier effective mass replaces the particle mass in mathematical relationships.*

The carrier effective masses appropriate for Si, Ge, and GaAs are listed in Table 2.1.*

---

*For the sake of clarity it is sometimes necessary to oversimplify or gloss over complicating issues. The effective mass discussion is a case in point. In reality, the m_n^* appearing in Eq. (2.3) is not a simple scalar quantity, as we have led the reader to believe; nor can one revamp all mathematical relationships using the carrier effective masses listed in Table 2.1. For any given material and carrier type, there are in fact several effective masses encountered in practice. The masses listed in Table 2.1, however, are the most commonly encountered effective masses. Moreover, the precise numerical values of the effective masses are really of secondary importance. Of prime importance is the drastic simplification resulting from the fact that the quantum-mechanical entities known as electrons and holes may be treated, both conceptually and mathematically, as classical particles.

Table 2.1 Density of States Effective Masses.

Material	m_n^*/m_0	m_p^*/m_0
Si	1.1	0.59
Ge	0.55	0.37
GaAs	0.068	0.54

2.3.3 Carrier Numbers in Intrinsic Material

The term "intrinsic semiconductor" in common usage refers to an extremely pure semiconductor sample containing an insignificant amount of impurity atoms. More precisely, an intrinsic semiconductor is a semiconductor whose properties are native to the material (that is, not caused by external additives). The number of carriers in an intrinsic semiconductor fits into the scheme of things in that it is an identifiable intrinsic property of the material.

Defining, quite generally,

$$n = \text{number of electrons/cm}^3,$$

$$p = \text{number of holes/cm}^3,$$

existing inside a semiconductor; then, given an intrinsic semiconductor under equilibrium conditions, one finds

$$n = p = n_i \tag{2.4}$$

and

$$n_i \simeq 10^{10}/\text{cm}^3 \quad \text{in Si at room temperature,}$$
$$\simeq 10^{13}/\text{cm}^3 \quad \text{in Ge at room temperature.}$$

Numbers to remember.

The electron and hole concentrations in an intrinsic semiconductor are equal because carriers within a very pure material can be created only in pairs. Referring to Fig. 2.7, if a semiconductor bond is broken, a free electron and a missing bond or hole are created simultaneously. Likewise, the excitation of an electron from the valence band into the conduction band automatically creates a valence band hole along with the conduction band electron. Also note that the intrinsic carrier concentration, although large in an absolute sense, is relatively small compared with the number of bonds that could be broken. For example, in Si there are 5×10^{22} atoms/cm^3 and four bonds per atom, making a grand total of 2×10^{23} bonds or valence band electrons per cm^3. Since $n_i \simeq 10^{10}/\text{cm}^3$, one finds less than one bond in 10^{13} broken in Si at room temperature. To accurately represent the situation inside intrinsic Si at room temperature, we could cover all the university chalkboards in the world with the bonding model and possibly show only *one* broken bond.

Table 2.2 Common Silicon Dopants. Arrows indicate the most widely employed dopants.

Donors (Electron-increasing Dopants)		Acceptors (Hole-increasing Dopants)	
P←	Column V elements	B←	Column III elements
As		Ga	
Sb		In	
		Al	

2.3.4 Manipulation of Carrier Numbers—Doping

Doping, in semiconductor terminology, is the addition of controlled amounts of specific impurity atoms with the expressed purpose of increasing either the electron or the hole concentration. The addition of dopants in controlled amounts to semiconductor materials, it should be pointed out, occurs routinely in the fabrication of almost all semiconductor devices. Common Si dopants are listed in Table 2.2. To increase the electron concentration, one can add either phosphorus, arsenic, or antimony atoms to the Si crystal, with phosphorus being the most commonly employed donor (electron-increasing) dopant. To increase the hole concentration, one adds either boron, gallium, indium, or aluminum atoms to the Si crystal, with boron being the most commonly employed acceptor (hole-increasing) dopant.

In order to understand how the addition of impurity atoms can lead to a manipulation of carrier numbers, it is important to note that the Table 2.2 donors are all from Column V in the Periodic Table of Elements, while all of the cited acceptors are from Column III in the Periodic Table of Elements. As visualized in Fig. 2.10(a) using the bonding model, where a Column V element with five valence electrons is substituted for a Si atom

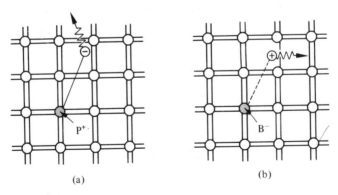

(a)   (b)

Fig. 2.10 Visualization of (a) donor and (b) acceptor action using the bonding model. In (a) the Column V element P is substituted for a Si atom; in (b) the Column III element B is substituted for a Si atom.

in the semiconductor lattice, four of the five valence electrons fit snugly into the bonding structure. The fifth donor electron, however, does not fit into the bonding structure, is weakly bound, and at room temperature is readily freed to wander about the lattice, and hence becomes a carrier. Please note that this donation (hence the name "donor") of carrier electrons does not increase the hole concentration. The P^+ (or other donor) ion left behind when the fifth electron is released cannot move, and there are no broken atom – atom bonds associated with the release of the fifth electron.

The explanation of acceptor action follows a similar line of reasoning. The Column III acceptors have three valence electrons and cannot complete one of the semiconductor bonds when substituted for Si atoms in the semiconductor lattice (see Fig. 2.10(b)). The Column III atom, however, readily accepts (hence the name "acceptor") an electron from a nearby Si – Si bond, thereby completing its own bonding scheme and in the process creating a hole that can wander about the lattice. Here again there is an increase in only one type of carrier. A negatively charged acceptor site (acceptor atom plus accepted electron) cannot move, and no electrons are released in the hole creation process.

The foregoing bonding-model-based explanation of dopant action is reasonably understandable. There are, nevertheless, a few loose ends. For one, we noted that the fifth donor electron was rather weakly bound and readily freed at room temperature. How does one interpret the relative term "weakly bound"? It takes $\simeq 1$ eV to break Si – Si bonds and very few of the Si – Si bonds are broken at room temperature. Perhaps "weakly bound" means a binding energy $\simeq 0.1$ eV or less? The question also arises as to how one visualizes dopant action in terms of the energy band model. Both of the cited questions, as it turns out, involve energy considerations and are actually interrelated.

Let us concentrate first on the binding energy of the fifth donor electron. Crudely speaking, the positively charged donor-core-plus-fifth-electron may be likened to a hydrogen atom (see Fig. 2.11). Conceptually, the donor core replaces the hydrogen-atom nucleus and the fifth donor electron replaces the hydrogen-atom electron. In the real hydrogen atom the electron moves of course in a vacuum, can be characterized by the mass of a free electron, and, referring to Eq. (2.1), has a ground-state binding energy of -13.6 eV. In the pseudo-hydrogen atom, on the other hand, the orbiting electron moves

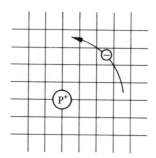

Fig. 2.11   Pseudo-hydrogen atom model for donor-site bonding.

through a sea of Si atoms and is characterized by an effective mass. Hence, in the donor or pseudo-atom case, the permittivity of free space must be replaced by the permittivity of Si and m_0 must be replaced by m_n^*. We therefore conclude that the binding energy (E_B) of the fifth donor electron is approximately

$$E_B \simeq -\frac{m_n^* q^4}{2(4\pi K_S \varepsilon_0 \hbar)^2} = \frac{m_n^*}{m_0} \frac{1}{K_S^2} E_{H|n=1} \simeq -0.1 \text{ eV} \tag{2.5}$$

where K_S is the Si dielectric constant ($K_S \simeq 12$). Actual donor-site binding energies in Si are listed in Table 2.3. The observed binding energies are seen to be in good agreement with the Eq. (2.5) estimate and, confirming the earlier speculation, are roughly $\simeq 1/20$ the Si band gap energy.

Having established the strength of dopant-site bonding, we are now in a position to tackle the problem of how one visualizes dopant action using the energy band model. Working on the problem, we note first of all that when an electron is released from a donor site it becomes a conduction band electron. If the energy absorbed at the donor site is precisely equal to the electron binding energy, the released electron will moreover have the lowest possible energy in the conduction band — namely, E_c. Adding an energy $|E_B|$ to the bound donor-site electron, in other words, *raises* the electron's energy to E_c. Hence, we are led to conclude that the bound electron occupies an allowed electronic level an energy $|E_B|$ below the conduction band edge, or, as visualized in Fig. 2.12, donor sites can be incorporated into the energy band scheme by adding allowed electronic levels at an energy $E_D = E_c - |E_B|$. Note that the donor-site energy level is represented by a set of dashes, instead of a continuous line, because an electron bound to a donor site is localized in space; i.e., a bound electron does not leave the general Δx vicinity of the donor site. The relative closeness of E_D to E_c of course reflects the fact that $E_c - E_D = |E_B| \simeq (1/20)E_G(\text{Si})$.

The actual visualization of dopant action using the energy band model is pictured in Fig. 2.13. Examining the left-hand side of Fig. 2.13(a), one finds all of the donor sites filled with bound electrons at temperatures $T \rightarrow 0°\text{K}$. This is true because very little thermal energy is available to excite electrons from the donor sites into the conduction band at these very low temperatures. The situation changes, of course, as the temperature

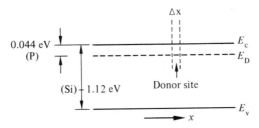

Fig. 2.12   Addition of the $E = E_D$ donor levels to the energy band diagram. Dashes of width Δx emphasize the localized nature of the bound donor-site states.

Table 2.3. Dopant-site Binding Energies.

Donors	$|E_B|$	Acceptors	$|E_B|$
Sb	0.039 eV	B	0.045 eV
P	0.044 eV	Al	0.057 eV
As	0.049 eV	Ga	0.065 eV
		In	0.16 eV

is increased, with more and more of the weakly bound electrons being donated to the conduction band. At room temperature the ionization of the donor sites is all but total, giving rise to the situation pictured at the extreme right of Fig. 2.13(a). Although we have concentrated on donors, the situation for acceptors is completely analogous. As visualized in Fig. 2.13(b), acceptors introduce allowed electronic levels into the forbidden gap at an energy slightly above the valence band edge. At low temperatures, all of these sites will be empty — there is insufficient energy at temperatures $T \rightarrow 0°K$ for a valence band electron to make the transition to an acceptor site. Increasing temperature, implying an increased store of thermal energy, facilitates electrons jumping from the valence band onto the acceptor levels. The removal of electrons from the valence band of course creates holes. At room temperature, essentially all of the acceptor sites are filled with an excess electron and an increased hole concentration is effected in the material.

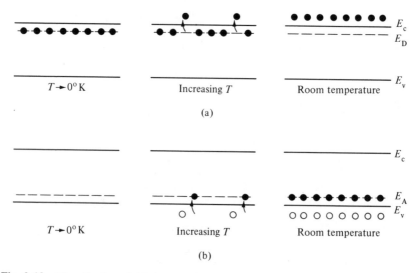

Fig. 2.13 Visualization of (a) donor and (b) acceptor action using the energy band model.

2.3.5 Carrier Related Terminology

Since terminology is often a stumbling block to understanding and since this particular section is replete with specialized terms, it seems appropriate to conclude the section with an overview of carrier related terminology. Approximately one half of the carrier related terms listed below were introduced and defined earlier in this section; the remainder of the terms are being listed here for the first time. All of the terms are widely employed and their definitions should be committed to memory.

Dopants — specific impurity atoms which are added to semiconductors in controlled amounts for the expressed purpose of increasing either the electron or the hole concentration.

Intrinsic semiconductor — undoped semiconductor; extremely pure semiconductor sample containing an insignificant amount of impurity atoms; a semiconductor whose properties are native to the material.

Extrinsic semiconductor — doped semiconductor; a semiconductor whose properties are controlled by added impurity atoms.

Donor — impurity atom which increases the electron concentration; n-type dopant.

Acceptor — impurity atom which increases the hole concentration; p-type dopant.

n-type material — a donor-doped material; a semiconductor containing more electrons than holes.

p-type material — an acceptor-doped material; a semiconductor containing more holes than electrons.

Majority carrier — the most abundant carrier in a given semiconductor sample; electrons in an n-type material, holes in a p-type material.

Minority carrier — the least abundant carrier in a given semiconductor sample; holes in an n-type material, electrons in a p-type material.

2.4 STATE AND CARRIER DISTRIBUTIONS

Up to this point in the modeling process we have concentrated on carrier properties and information of a conceptual, qualitative or, at most, semiquantitative nature. Practically speaking, there is often a need for more detailed information. For example, most semiconductors are doped, and the precise numerical value of the carrier concentrations inside doped semiconductors is of routine interest. Another property of interest is the distribution of carriers as a function of energy in the respective energy bands. In this section we begin the process of developing a more detailed description of the carrier populations. The development will eventually lead to relationships for the carrier distributions and concentrations within semiconductors under equilibrium conditions.

2.4.1 Density of States

When the energy band model was first introduced in Section 2.2 we indicated that the total number of allowed states in each band was four times the number of atoms in the crystal. Not mentioned at the time was how the allowed states were distributed in energy; that is, how many states were to be found at any given energy in the conduction and valence bands. We are now interested in this energy distribution of states, or *density of states*, as it is more commonly known, because the state distribution is an essential component in determining carrier distributions and concentrations.

To determine the desired density of states it is necessary to perform an analysis based on quantum-mechanical considerations. Herein we will merely summarize the results of the analysis; namely, for energies not too far removed from the band edges, one finds

$$g_c(E) = \frac{m_n^* \sqrt{2m_n^*(E - E_c)}}{\pi^2 \hbar^3}, \qquad E \geq E_c \qquad (2.6a)$$

$$g_v(E) = \frac{m_p^* \sqrt{2m_p^*(E_v - E)}}{\pi^2 \hbar^3}, \qquad E \leq E_v \qquad (2.6b)$$

where $g_c(E)$ and $g_v(E)$ are the density of states at an energy E in the conduction and valence bands, respectively.

What exactly should the reader know and remember about the cited density of states? Well, for one, it is important to grasp the general density of states concept. The density of states can be likened to the description of the seating in a football stadium, with the number of seats in the stadium a given distance from the playing field corresponding to the number of states a specified energy interval from E_c or E_v. Secondly, the knowledgeable reader would remember the general form of the relationships. As illustrated in Fig. 2.14, $g_c(E)$ is zero at E_c and increases as the square root of energy when one proceeds upward into the conduction band. Similarly, $g_v(E)$ is precisely zero at E_v and increases with the square root of energy as one proceeds downward from E_v into the valence band. Thirdly, one should note that differences between $g_c(E)$ and $g_v(E)$ stem from differences in the carrier effective masses. If m_n^* were equal to m_p^*, the seating (states) on both sides of the football field (the band gap) would be mirror images of each other. Finally, considering closely spaced energies E and $E + dE$ in the respective bands, one can state

$g_c(E)dE$ represents the number of conduction band states/cm^3 lying in the energy range between E and $E + dE$ (if $E \geq E_c$),

$g_v(E)dE$ represents the number of valence band states/cm^3 lying in the energy range between E and $E + dE$ (if $E \leq E_v$).

It therefore follows that $g_c(E)$ and $g_v(E)$ themselves are numbers/unit volume-unit energy, or typically, numbers/cm^3 − eV.

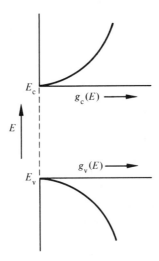

Fig. 2.14 General energy dependence of $g_c(E)$ and $g_v(E)$ near the band edges. $g_c(E)$ and $g_v(E)$ are the density of states in the conduction and valence bands, respectively.

2.4.2 The Fermi Function

Whereas the density of states tells one how many states exist at a given energy E, the Fermi function $f(E)$ specifies how many of the existing states at the energy E will be filled with an electron, or equivalently,

    $f(E)$ specifies, under equilibrium conditions, the probability that an available state at an energy E will be occupied by an electron.

Mathematically speaking, the Fermi function is simply a probability density function. In mathematical symbols,

$$f(E) = \frac{1}{1 + e^{(E-E_F)/kT}} \tag{2.7}$$

where

    E_F = Fermi energy or Fermi level

    k = Boltzmann constant ($k = 8.62 \times 10^{-5} \text{eV}/°\text{K}$)

    T = temperature in $°\text{K}$

    Seeking insight into the nature of the newly introduced function, let us begin by investigating the Fermi function's energy dependence. Consider first temperatures where $T \to 0°\text{K}$. As $T \to 0°\text{K}$, $(E - E_F)/kT \to -\infty$ for all energies $E < E_F$ and $(E - E_F)/kT \to +\infty$ for all energies $E > E_F$. Hence $f(E < E_F) \to 1/[1 + \exp(-\infty)] = 1$ and $f(E > E_F) \to 1/[1 + \exp(+\infty)] = 0$. This result is plotted in Fig. 2.15(a) and is simply

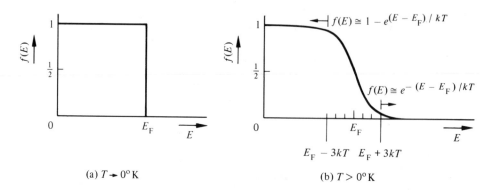

Fig. 2.15 Energy dependence of the Fermi function. (a) $T \rightarrow 0°$K; (b) generalized $T > 0°$K plot with the energy coordinate expressed in kT units.

interpreted as meaning that all states at energies below E_F will be filled and all states at energies above E_F will be empty for temperatures $T \rightarrow 0°$K. In other words, there is a sharp cutoff in the filling of allowed energy states at the Fermi energy E_F when the system temperature approaches absolute zero.

Let us next consider temperatures $T > 0°$K. Examining the Fermi function we make the following pertinent observations.

i) If $E = E_F$, $f(E_F) = 1/2$.

ii) If $E \geq E_F + 3kT$, $\exp[(E-E_F)/kT] \gg 1$ and $f(E) \simeq \exp[-(E-E_F)/kT]$. Consequently, above $E_F + 3kT$ the Fermi function or filled-state probability decays exponentially to zero with increasing energy. Moreover, most states at energies $3kT$ or more above E_F will be empty.

iii) If $E \leq E_F - 3kT$, $\exp[(E-E_F)/kT] \ll 1$ and $f(E) \simeq 1 - \exp[(E-E_F)/kT]$. Below $E_F - 3kT$, therefore, $[1 - f(E)]$, the probability that a given state will be *empty*, decays exponentially to zero with decreasing energy. Most states at energies $3kT$ or more below E_F will be filled.

iv) At room temperature ($T \simeq 300°$K), $kT \simeq 0.026$ eV and $3kT \simeq 0.078$ eV $\ll E_G$(Si). Compared to the Si band gap, the $3kT$ energy interval that appears prominently in the $T > 0°$K formalism is typically quite small.

The properties cited above are reflected and summarized in the $T > 0°$K Fermi function plot displayed in Fig. 2.15(b).

Before concluding the discussion here, it is perhaps worthwhile to reemphasize that the Fermi function applies only under equilibrium conditions. The Fermi function, however, is universal in the sense it applies with equal validity to all materials —insulators, semiconductors, and metals. Although introduced in relationship to semiconductors, the Fermi function is not dependent in any way on the special nature of semiconductors, but is simply a statistical function associated with electrons in general.

Finally, the relative positioning of the Fermi energy E_F compared to E_c (or E_v), an item of obvious concern, is treated in subsequent subsections.

2.4.3 Equilibrium Distribution of Carriers

Having established the distribution of available band states and the probability of filling those states under equilibrium conditions, we can now easily deduce the distribution of carriers in the respective energy bands. To be specific, the desired distribution is obtained by simply multiplying the appropriate density of states by the appropriate occupancy factor—$f(E)g_c(E)$ yields the distribution of electrons in the conduction band and $[1 - f(E)]g_v(E)$ yields the distribution of holes (unfilled states) in the valence band. Sample carrier distributions for three different assumed positions of the Fermi energy (along with associated energy band diagram, Fermi function, and density of states plots) are pictured in Fig. 2.16.

Examining Fig. 2.16 we note in general that all carrier distributions are zero at the band edges, reach a peak value very close to E_c or E_v, and then decay very rapidly toward zero as one moves upward into the conduction band or downward into the valence band. In other words, most of the carriers are grouped energetically in the near vicinity of the band edges. Another general point to note is the effect of the Fermi level positioning on the relative magnitude of the carrier distributions. When E_F is positioned in the upper half of the band gap (or higher), the electron distribution greatly outweighs the hole distribution. Although both the filled-state occupancy factor $[f(E)]$ and the empty-state occupancy factor $[1 - f(E)]$ fall off exponentially as one proceeds from the band edges deeper into the conduction and valence bands, respectively, $[1 - f(E)]$ is clearly much smaller than $f(E)$ at corresponding energies if E_F lies in the upper half of the band gap. Lowering E_F effectively slides the occupancy plots downward, giving rise to a nearly equal number of carriers when E_F is at the middle of the gap, and a predominance of holes when E_F lies below the middle of the gap. The argument here assumes, of course, that $g_c(E)$ and $g_v(E)$ are of the same order of magnitude at corresponding energies (the usual case). Also, referring back to the previous subsection, the statements concerning the occupancy factors falling off exponentially in the respective bands are valid provided $E_c - 3kT \geq E_F \geq E_v + 3kT$.

The information just presented concerning the carrier distributions and the relative magnitudes of the carrier numbers finds widespread usage. The information, however, is often conveyed in an abbreviated or shorthand fashion. Figure 2.17, for example, shows a common way of representing the carrier energy distributions. The greatest number of circles or dots are drawn close to E_c and E_v, reflecting the peak in the carrier concentrations near the band edges. The smaller number of dots as one progresses upward into the conduction band crudely models the rapid falloff in the electron density with increasing energy. An extensively utilized means of conveying the relative magnitude of the carrier numbers is displayed in Fig. 2.18. To represent an intrinsic material, a dashed line is drawn in approximately the middle of the band gap and labeled E_i. The near-midgap positioning of E_i, the intrinsic Fermi level, is of course consistent with the previously cited fact that

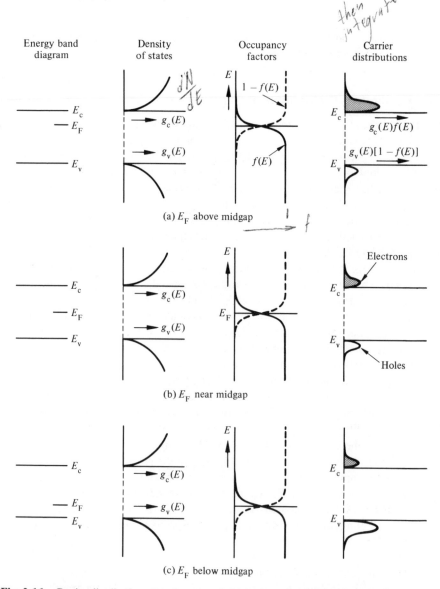

Fig. 2.16 Carrier distributions (not drawn to scale) in the respective bands when the Fermi level is positioned (a) above midgap, (b) near midgap, and (c) below midgap. Also shown in each case are coordinated sketches of the energy band diagram, density of states, and the occupancy factors (the Fermi function and one minus the Fermi function).

the electron and hole numbers are about equal when E_F is near the center of the band gap. Similarly, a solid line labeled E_F appearing above midgap tells one at a glance that the semiconductor in question is n-type; a solid line labeled E_F appearing below midgap

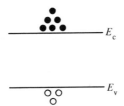

Fig. 2.17   Schematic representation of carrier energy distributions.

signifies that the semiconductor is p-type. Note finally that the dashed E_i line also typically appears on the energy band diagrams characterizing extrinsic semiconductors. The E_i line in such cases represents the expected positioning of the Fermi level *if* the material were intrinsic, and serves as a reference energy level dividing the upper and lower halves of the band gap.

2.5   EQUILIBRIUM CARRIER CONCENTRATIONS

We have arrived at a somewhat critical point in the carrier modeling process. For the most part, this section simply embodies the culmination of our modeling efforts, with working relationships for the equilibrium carrier concentrations being established to reinforce the qualitative carrier information presented in previous sections. Unfortunately, the emphasis on the development of mathematical relationships makes the final assault on the carrier modeling summit the most difficult part of the journey — the reader cannot relax. A comment is also in order concerning the presentation herein of alternative forms for the carrier relationships. The alternative forms can be likened to the different kinds of wrenches used, for example, in home and automobile repairs. The open-end wrench, the box wrench, and the ratchet wrench all serve the same general purpose. In some applications one can use any of the wrenches. In other applications, however, a special situation restricts the type of wrench employed or favors the use of one wrench over another. The same is true of the alternative carrier relationships. Finally, expressions that are widely quoted or find widespread usage have been enclosed in brackets. A single set of brackets signifies a moderately important result; a double set of brackets, a very important result.

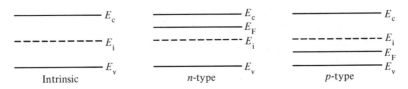

Fig. 2.18   "At a glance" representation of intrinsic (left), n-type (middle), and p-type (right) semiconductor materials using the energy band diagram.

2.5.1   Formulas for *n* and *p*

Since $g_c(E)\,dE$ represents the number of conduction band states/cm^3 lying in the E to $E + dE$ energy range, and $f(E)$ specifies the probability that an available state at an energy E will be occupied by an electron, it then follows that $f(E)g_c(E)dE$ gives the number of conduction band electrons/cm^3 lying in the E to $E + dE$ energy range, and $f(E)g_c(E)dE$ integrated over all conduction band energies must yield the total number of electrons in the conduction band. In other words, integration over the equilibrium distribution of electrons in the conduction band yields the equilibrium electron concentration. A similar statement can be made relative to the hole concentration. We therefore conclude

$$n = \int_{E_c}^{E_{top}} f(E)g_c(E)\,dE \tag{2.8a}$$

$$p = \int_{E_{bottom}}^{E_v} [1 - f(E)]\,g_v(E)\,dE \tag{2.8b}$$

Seeking explicit expressions for the carrier concentrations, let us focus our efforts on the *n*-integral. (It will be assumed the reader can perform the *p*-integral manipulations following an analogous procedure.) Substituting the Eq. (2.6a) expression for $g_c(E)$ and the Eq. (2.7) expression for $f(E)$ into Eq. (2.8a), one obtains

$$n = \frac{m_n^* \sqrt{2m_n^*}}{\pi^2 \hbar^3} \int_{E_c}^{E_{top}} \frac{\sqrt{E - E_c}\,dE}{1 + e^{(E-E_F)/kT}} \tag{2.9}$$

Now letting

$$\eta = \frac{(E - E_c)}{kT} \tag{2.10a}$$

$$\eta_f = \frac{(E_F - E_c)}{kT} \tag{2.10b}$$

$$E_{top} \rightarrow \infty \tag{2.10c}$$

yields

$$n = \frac{m_n^* \sqrt{2m_n^*}\,(kT)^{3/2}}{\pi^2 \hbar^3} \int_0^\infty \frac{\eta^{1/2}d\eta}{1 + e^{\eta - \eta_f}} \tag{2.11}$$

The (2.10c) simplification on the upper integration limit makes use of the fact that the integrand in question falls off rapidly with increasing energy and is essentially zero for energies only a few kT above E_c. Hence, extending the upper limit to ∞ has a totally negligible effect on the value of the integral.

Even with the cited simplification, the Eq.(2.11) integral cannot, in general, be expressed in a closed form containing simple functions. Faced with a dilemma of this type one does the next best thing — one associates a name and special symbol with the integral. Identifying

$$F_{1/2}(\eta_f) \equiv \int_0^\infty \frac{\eta^{1/2} d\eta}{1 + e^{\eta - \eta_f}}, \qquad \begin{array}{c} \text{the Fermi integral of order } \tfrac{1}{2} \\ \text{(a tabulated function)} \end{array} \qquad (2.12)$$

and also defining

$$N_C = 2 \left[\frac{2\pi m_n^* kT}{h^2} \right]^{3/2}, \qquad \begin{array}{c} \text{the "effective" density of} \\ \text{conduction band states} \end{array} \qquad (2.13)$$

one obtains

$$\boxed{n = N_C \frac{2}{\sqrt{\pi}} F_{1/2}(\eta_f)} \qquad (2.14a)$$

and, by analogy,

$$\boxed{p = N_V \frac{2}{\sqrt{\pi}} F_{1/2}(\eta_f')} \qquad (2.14b)$$

where $N_C \rightarrow N_V$ if $m_n^* \rightarrow m_p^*$ and $\eta_f' \equiv (E_v - E_F)/kT$.

The Eq.(2.14) relationships are a very general result, valid for any conceivable positioning of the Fermi level. The constants N_C and N_V are readily calculated and the value of the Fermi integral can be obtained from available tables, plots, or by direct computation. The general-form relationships, nonetheless, are admittedly cumbersome and inconvenient to use in routine analyses. Fortunately, simplified closed-form expressions do exist which can be employed in the vast majority of practical problems. To be specific, if E_F is restricted to values $E_F \leq E_c - 3kT$, then $1/[1 + \exp(\eta - \eta_f)] \simeq \exp[-(\eta - \eta_f)]$ for all $E \geq E_c$ ($\eta \geq 0$), and

$$F_{1/2}(\eta_f) = \frac{\sqrt{\pi}}{2} e^{(E_F - E_c)/kT} \qquad (2.15a)$$

Likewise, if $E_F \geq E_v + 3kT$, then

$$F_{1/2}(\eta_f') = \frac{\sqrt{\pi}}{2} e^{(E_v - E_F)/kT} \qquad (2.15b)$$

It therefore follows that, if $E_v + 3kT \leq E_F \leq E_c - 3kT$,

$$\boxed{\begin{array}{l} n = N_C e^{(E_F - E_c)/kT} \\ p = N_V e^{(E_v - E_F)/kT} \end{array}} \qquad \begin{array}{l} (2.16a) \\ (2.16b) \end{array}$$

The alert reader will recognize the mathematical simplification leading to Eqs. (2.16) as being identical to approximating the filled-state and empty-state occupancy factors by exponential functions — an approximation earlier shown to be valid provided E_F was somewhere in the band gap no closer than $3kT$ to either band edge. Whenever E_F is confined, as noted, to $E_v + 3kT \le E_F \le E_c - 3kT$, instead of continually repeating the E_F restriction, the semiconductor is simply said to be *nondegenerate*. Whenever E_F lies in the band gap closer than $3kT$ to either band edge or actually penetrates one of the bands, the semiconductor is said to be *degenerate*. These very important terms are also defined pictorially in Fig. 2.19.

2.5.2 Alternative Expressions for n and p

Although in closed form, the Eq. (2.16) relationships are not in the simplest form possible, and, more often than not, it is the simpler alternative form of these relationships which one encounters in device analyses. The alternative-form relationships can be obtained by recalling that E_i, the Fermi Level for an intrinsic semiconductor, lies close to midgap, and hence Eqs. (2.16) most assuredly apply to an intrinsic semiconductor. If this be the case, then specializing Eqs. (2.16) to an intrinsic semiconductor, i.e., setting $n = p = n_i$ and $E_i = E_F$, one obtains

$$n_i = N_C e^{(E_i - E_c)/kT} \tag{2.17a}$$

and

$$n_i = N_V e^{(E_v - E_i)/kT} \tag{2.17b}$$

Rearranging Eqs. (2.17) yields

$$N_C e^{-E_c/kT} = n_i e^{-E_i/kT} \tag{2.18a}$$

and

$$N_V e^{E_v/kT} = n_i e^{E_i/kT} \tag{2.18b}$$

Finally, eliminating $N_C \exp(-E_c/kT)$ and $N_V \exp(E_v/kT)$ in the original Eq. (2.16) relationships using Eqs. (2.18) gives

$$\left[\begin{array}{l} n = n_i e^{(E_F - E_i)/kT} \\ p = n_i e^{(E_i - E_F)/kT} \end{array} \right] \tag{2.19a}$$
$$\tag{2.19b}$$

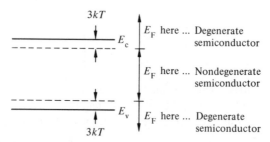

Fig. 2.19 Definition of degenerate/nondegenerate semiconductors.

Like Eq. (2.16), the Eq. (2.19) expressions are valid for any semiconductor in equilibrium whose doping is such as to give rise to a nondegenerate positioning of the Fermi level. However, whereas two constants and three energy levels appear in the original relationships, only one constant and two energy levels appear in the alternative relationships. Because of their symmetrical nature, the alternative expressions are also easier to remember, requiring merely an interchange of E_F and E_i in going from the n-formula to the p-formula.

2.5.3 n_i and the np Product

Continuing to establish pertinent carrier concentration relationships, let us next briefly interject considerations relevant to the intrinsic carrier concentration. If the corresponding sides of Eqs. (2.17a) and (2.17b) are multiplied together, one obtains

$$n_i^2 = N_C N_V e^{-(E_c - E_v)/kT} = N_C N_V e^{-E_G/kT} \tag{2.20}$$

or, effecting the indicated square root,

$$\left[n_i = \sqrt{N_C N_V}\, e^{-E_G/2kT} \right] \tag{2.21}$$

Equation (2.21) expresses n_i as a function of known quantities and may be used to compute n_i at a specified temperature or as a function of temperature. Numerical values for the intrinsic carrier concentration in Si and Ge at room temperature were cited previously; the best available plots of n_i as functions of temperature in Si, Ge, and GaAs are displayed in Fig. 2.20.*

Another very important n_i-based carrier relationship can be readily extracted from Eqs. (2.19). If the corresponding sides of Eqs. (2.19a) and (2.19b) are multiplied together, one rapidly establishes

$$\llbracket np = n_i^2 \rrbracket \tag{2.22}$$

Seemingly trivial, Eq. (2.22) leads immediately to several interesting conclusions. Whereas doping a semiconductor increases either the electron or hole concentration, the np product, as deduced from Eq. (2.22), remains unperturbed, provided of course that the doped semiconductor is in equilibrium and nondegenerate. Moreover, a doping-related increase in the electron concentration must be accompanied by a corresponding decrease in the hole concentration and vice versa. Also note that, given one of the carrier concentrations, the remaining carrier concentration is easily determined using Eq. (2.22).

2.5.4 Charge Neutrality

The charge neutrality relationship will be our final addition to the rather impressive arsenal of fundamental concentration-related expressions. To establish the charge neutrality

---

*Substitution of the parameters provided herein into Eq. (2.21) will not yield n_i values precisely as graphed in Fig. 2.20. The primary reason for this discrepancy is the weak but nonnegligible temperature dependence of the effective masses and E_G.

relationship, let us consider a *uniformly doped* semiconductor, a semiconductor where the number of dopant atoms/cm^3 is the same everywhere. Systematically examining little sections of the semiconductor and assuming equilibrium conditions, one must invariably find that each and every section of the material is charge-neutral, that is, it exhibits no net charge. If this were not the case, electric fields would exist inside the semiconductor which in turn would give rise to carrier motion and associated currents — a situation totally inconsistent with the assumed equilibrium conditions. There are, however, charged entities inside all semiconductors. Electrons, holes, donor sites that have donated an electron to the conduction band (referred to as ionized donor sites), and ionized (negatively charged) acceptor sites may all exist simultaneously inside any given semiconductor. For the uniformly doped material to be everywhere charge-neutral clearly requires

$$\frac{\text{charge}}{\text{cm}^3} = qp - qn + qN_D^+ - qN_A^- = 0 \tag{2.23}$$

or

$$[p - n + N_D^+ - N_A^- = 0] \tag{2.24}$$

where, by definition,

N_D^+ = number of ionized (positively charged) donor sites/cm^3,

N_A^- = number of ionized (negatively charged) acceptor sites/cm^3.

As previously discussed, there is sufficient thermal energy available in a semiconductor at room temperature to ionize almost all of the shallow-level donor and acceptor sites. Setting

$$N_D^+ = N_D$$
$$N_A^- = N_A$$

where

N_D = total number of donor atoms or sites/cm^3,

N_A = total number of acceptor atoms or sites/cm^3,

one then obtains

$$[\![p - n + N_D - N_A = 0]\!] \qquad \begin{array}{c} \text{assumes total ionization} \\ \text{of dopant sites} \end{array} \tag{2.25}$$

Equation (2.25) is the standard form of the charge neutrality relationship.

2.5.5 Carrier Concentration Calculations

Armed with the standard arsenal of fundamental carrier relationships, we are now in a position to survey example uses of the formalism and, whenever possible, to establish additional carrier related facts and information. Let us begin by briefly reexamining the

np product and charge neutrality relationships. n_i, which appears in the np product expression, has been calculated and plotted and must be considered a known quantity. Likewise, N_A and N_D, which appear in the charge neutrality relationship, are typically controlled and determined experimentally, and should also be considered known quantities. The only other symbols used in the two equations are n and p. Clearly, we have two equations and two unknowns from which n and p can be deduced.

(1)   *Intrinsic Semiconductor* $(N_D = 0, N_A = 0)$. For our initial calculation, as a sort of test run, we tackle a problem with a known solution. Setting $N_D = 0$ and $N_A = 0$,

$$p - n + N_D - N_A = 0 \rightarrow p = n \tag{2.26}$$

Next substituting $p = n$ into the np product expression,

$$np = n^2 = p^2 = n_i^2 \rightarrow n = p = n_i \tag{2.27}$$

$n = p = n_i$ is, of course, the expected result for the equilibrium carrier concentration in an intrinsic semiconductor.

(2)   *Extrinsic Semiconductor (nondegenerate, fully ionized).* Since most devices are made from doped semiconductors and are operated at or above room temperature, the case at hand is of obvious practical interest. The only general computational assumptions are those of equilibrium, nondegeneracy (allowing us to use the np product expression), and total ionization of dopant sites. In the sample computation we will also assume a donor doped (n-type) material where $N_D \gg N_A$, or $N_D - N_A \simeq N_D$. The computation for an acceptor doped (p-type) material follows by analogy.

Starting with the charge neutrality relationship appropriate for a donor doped semiconductor, we can write

$$p - n + N_D = 0 \tag{2.28}$$

However, from the np product expression

$$p = \frac{n_i^2}{n} \tag{2.29}$$

Eliminating p in Eq. (2.28) using Eq. (2.29), one obtains

$$\frac{n_i^2}{n} - n + N_D = 0 \tag{2.30}$$

or

$$n^2 - nN_D - n_i^2 = 0 \tag{2.31}$$

Solving the quadratic equation for n yields

$$n = \frac{N_D}{2} + \left[\left(\frac{N_D}{2} \right)^2 + n_i^2 \right]^{1/2} \tag{2.32}$$

Only the plus root has been retained because physically we must have $n \geq 0$.

For a given donor doping and ambient temperature (or n_i), Eq. (2.32) specifies n; once n is known, Eq. (2.29) may be employed to compute p. Consider, however, that at room temperature n_i for Si is about $10^{10}/cm^3$ while doping concentrations are seldom less than $10^{14}/cm^3$, or typically, $N_D \gg n_i$. With $N_D \gg n_i$, the square root in Eq. (2.32) reduces to $N_D/2$, $n \simeq N_D$, and $p \simeq n_i^2/N_D$. An analogous statement can be made for an acceptor doped material where $N_A \gg N_D$; i.e., with $N_A \gg n_i$, $p \simeq N_A$ and $n \simeq n_i^2/N_A$. Repeating,

$$\left[\begin{array}{l} n = N_D \\ p = \dfrac{n_i^2}{N_D} \end{array}\right] \qquad \begin{array}{ll} N_D \gg N_A; & N_D \gg n_i \\ \text{(nondegenerate; fully ionized)} \end{array} \qquad \begin{array}{l} (2.33a) \\ \\ (2.33b) \end{array}$$

$$\left[\begin{array}{l} p = N_A \\ n = \dfrac{n_i^2}{N_A} \end{array}\right] \qquad \begin{array}{ll} N_A \gg N_D; & N_A \gg n_i \\ \text{(nondegenerate; fully ionized)} \end{array} \qquad \begin{array}{l} (2.34a) \\ \\ (2.34b) \end{array}$$

If any of the equations contained in this section are committed to memory, Eqs. (2.33) and (2.34) should be at the top of the list. Under typical conditions, it is a good bet these equations will be used to compute the equilibrium carrier concentrations. In fact, *unless a statement is made to the contrary or conditions indicate otherwise, one should automatically assume that the equilibrium carrier concentrations can be deduced from either Eq. (2.33) or Eq. (2.34).*

One final point: As the ambient temperature increases, n_i becomes larger and larger (see Fig. 2.20), and will eventually equal and then exceed N_D (or N_A). At elevated temperatures where $n_i \gg N_D$, the square root in Eq. (2.32) reduces to n_i, $n \simeq n_i$, and $p \simeq n_i$. In other words, all semiconductors become intrinsic at sufficiently high temperatures.

2.5.6 Finding the Fermi Level

Knowledge concerning the exact position of the Fermi level on the energy band diagram is often of interest. For example, when discussing the intrinsic Fermi level we indicated that E_i was located somewhere near the middle of the band gap. It would be useful to know the *precise* positioning of E_i in the band gap. Moreover, we have developed computational formulas for n and p appropriate for nondegenerate semiconductors. Whether a doped semiconductor is nondegenerate or degenerate depends, of course, on the value or positioning of E_F.

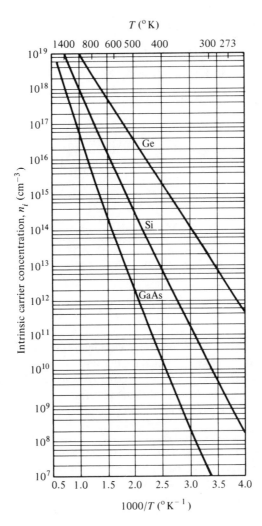

Fig. 2.20   Temperature dependence of the intrinsic carrier concentrations in Si, Ge, and GaAs. (From *Physics of Semiconductor Devices* by S. M. Sze, ©1969 by John Wiley and Sons, Inc.)

Before running through the mechanics of finding the Fermi level in selected cases of interest, it is useful to make a general observation; namely, Eqs. (2.19) or (2.16) [or even more generally, Eqs. (2.14)] provide a one-to-one correspondence between the Fermi energy and the carrier concentrations. Having computed any one of the three variables — n, p, or E_F — one can always determine the remaining two variables under equilibrium conditions.

(1) Exact positioning of E_i. In an intrinsic material

$$n = p \tag{2.35}$$

Substituting for n and p in Eq. (2.35) using Eq. (2.16), and setting $E_F = E_i$, yields

$$N_C e^{(E_i - E_C)/kT} = N_V e^{(E_V - E_i)/kT} \tag{2.36}$$

Solving for E_i, one obtains

$$E_i = \frac{E_c + E_v}{2} + \frac{kT}{2} \ln\left(\frac{N_V}{N_C}\right) \tag{2.37}$$

But

$$\frac{N_V}{N_C} = \left(\frac{m_p^*}{m_n^*}\right)^{3/2} \tag{2.38}$$

Consequently,

$$\left[E_i = \frac{E_c + E_v}{2} + \frac{3}{4}kT \ln\left(\frac{m_p^*}{m_n^*}\right) \right] \tag{2.39}$$

According to Eq. (2.39), E_i lies precisely at midgap only if $m_p^* = m_n^*$ or if $T = 0°K$. For the more practical case of silicon at room temperature, Table 2.1 gives $m_p^*/m_n^* = 0.54$, $(3/4)kT \ln(m_p^*/m_n^*) = -0.012$ eV, and E_i therefore lies 0.012 eV below midgap. Although potentially significant in certain problems, this small deviation from midgap is typically neglected in drawing energy band diagrams, etc.

(2) Extrinsic Semiconductors (nondegenerate, fully ionized). The first order of business here is to develop formulas for the positioning of the Fermi level in donor and acceptor doped semiconductors assumed to be nondegenerate, in equilibrium, and maintained at or near room temperature. As established previously, in a typical $(N_D \gg N_A, N_D \gg n_i)$ donor-doped semiconductor,

$$n = N_D \tag{2.33a}$$

Eliminating n in Eq. (2.33a) using Eq. (2.19a) gives

$$n_i e^{(E_F - E_i)/kT} = N_D \tag{2.40}$$

Solving Eq. (2.40) for $E_F - E_i$ rapidly yields

$$[E_F - E_i = kT \ln(N_D/n_i)] \qquad n\text{-type material} \tag{2.41}$$

Similarly, for a typical $(N_A \gg N_D, N_A \gg n_i)$ acceptor-doped semiconductor maintained at or near room temperature,

$$p = N_A \tag{2.34a}$$

Thus, making use of Eq. (2.19b),

$$n_i e^{(E_i - E_F)/kT} = N_A \tag{2.42}$$

and

$$\left[E_i - E_F = kT \, \ln\!\left(\frac{N_A}{n_i}\right) \right] \qquad p\text{-type material} \tag{2.43}$$

From Eqs. (2.41) and (2.43) it is obvious that the Fermi level moves systematically upward in energy from E_i with increasing donor doping and systematically downward in energy from E_i with increasing acceptor doping. The exact Fermi level positioning in Si at room temperature, nicely reinforcing the foregoing statement, is displayed in Fig. 2.21. Also note that for a given semiconductor material and ambient temperature there exist maximum nondegenerate donor and acceptor concentrations, doping concentrations above which the material becomes degenerate. In Si at room temperature the maximum nondegenerate doping concentrations are $N_D \simeq 1.6 \times 10^{18}/\text{cm}^3$ and $N_A \simeq 7.7 \times 10^{17}/\text{cm}^3$. The large Si doping values required for degeneracy, we should interject, have led to the common usage of "highly doped" (or n^+-material/p^+-material) and "degenerate" as interchangeable terms. Finally, the question may arise, "What procedure should be followed in computing E_F when one is not sure whether a material

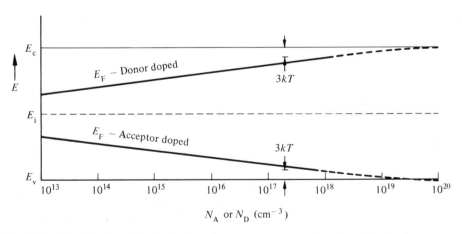

Fig. 2.21 Fermi level positioning in Si at room temperature as a function of the doping concentration. The solid E_F lines were established using Eq. (2.41) for donor doped material and Eq. (2.43) for acceptor doped material ($kT = 0.026$ eV, and $n_i = 10^{10}/\text{cm}^3$).

is nondegenerate or degenerate?" Unless a material is known to be degenerate, always assume nondegeneracy and compute E_F employing the appropriate nondegenerate relationship. If E_F derived from the nondegenerate formula lies in the degenerate zone, one must then, of course, recompute E_F using the more complex formalism valid for degenerate materials.

2.5.7 Carrier Concentration Temperature Dependence

A number of isolated facts about the carrier concentration temperature dependence have already been presented at various points in this chapter. The Section 2.3 discussion concerned with dopant action, for example, described the increased ionization of dopant sites and the associated increase in the majority carrier concentration when the temperature of a semiconductor is raised from near $T = 0°K$ toward room temperature. More recent subsections have included a plot of the intrinsic carrier concentration versus temperature (Fig. 2.20) and a calculation indicating that all semiconductors become intrinsic ($n \rightarrow n_i$, $p \rightarrow n_i$) at sufficiently high temperatures. In this subsection, which concludes the carrier concentration discussion, temperature-related facts are combined and embellished to provide a broader, more complete description of how the carrier concentrations vary with temperature.

Figure 2.22(a), a typical majority-carrier concentration-versus-temperature plot constructed assuming a phosphorus-doped $N_D = 10^{15}/cm^3$ Si sample, nicely illustrates the general features of the concentration-versus-temperature dependence. Examining Fig. 2.22(a) we find that n is fixed at approximately N_D over a broad temperature range extending roughly from 150°K to 450°K for the given Si sample. This $n \simeq N_D$ or "extrinsic temperature region" constitutes the normal operating range for most solid state devices. Below 100°K or so, in the "freeze-out temperature region," n drops significantly below N_D and approaches zero as $T \rightarrow 0°K$. In the "intrinsic temperature region" at the opposite end of the temperature scale, n rises above N_D, asymptotically approaching n_i with increasing T.

To qualitatively explain the just-described concentration-versus-temperature dependence, it is important to recall that the equilibrium number of carriers within a material is affected by two separate mechanisms. Electrons donated to the conduction band from donor sites and valence band electrons excited across the band gap into the conduction band (broken Si—Si bonds) both contribute to the majority-carrier electron concentration in a donor-doped material. At temperatures $T \rightarrow 0°K$ the thermal energy available in the system is insufficient to release the weakly bound fifth electron on donor sites and totally insufficient to excite electrons across the band gap. Hence $n = 0$ at $T = 0°K$, as visualized on the left-hand side of Fig. 2.22(b). Slightly increasing the material temperature above $T = 0°K$ "defrosts" or frees some of the electrons weakly bound to donor sites. Band-to-band excitation, however, remains extremely unlikely, and therefore the number of observed electrons in the freeze-out temperature region equals the number of ionized donor sites — $n = N_D^+$. Continuing to increase the system temperature eventually frees almost all of the weakly bound electrons on donor sites, n approaches N_D, and one

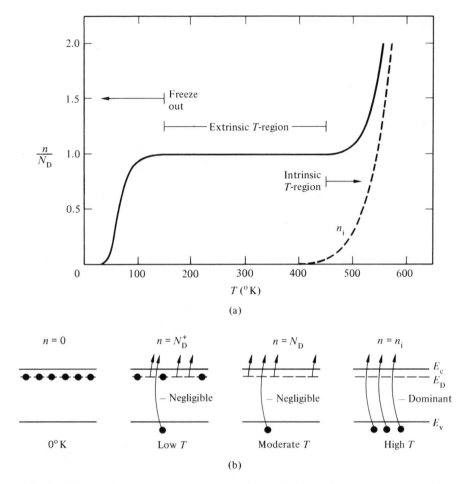

Fig. 2.22 (a) Typical temperature dependence of the majority carrier concentration in a doped semiconductor. The plot was constructed assuming a phosphorus-doped $N_D = 10^{15}/cm^3$ Si sample. n_i versus T (dashed line) has been included for comparison purposes. (b) Qualitative explanation of the concentration-versus-temperature dependence displayed in part (a).

enters the extrinsic temperature region. In progressing through the extrinsic temperature region, more and more electrons are excited across the band gap, but the number of electrons supplied in this fashion stays comfortably below N_D. Ultimately, of course, electrons excited across the band gap equal, then exceed and, as pictured on the right-hand side of Fig. 2.22(b), finally swamp the fixed number of electrons derived from the donor sites.

As a practical note, it should be pointed out that the wider the band gap, the greater the energy required to excite electrons from the valence band into the conduction band,

and the higher the temperature at the onset of the intrinsic temperature region. For this reason, Si devices can operate at higher temperatures than Ge devices, $\simeq 200°C$ at maximum for Si devices versus $\simeq 100°C$ at maximum for Ge devices. GaAs devices, of course, can operate at even higher temperatures than Si devices.

2.6 SUMMARY AND CONCLUDING COMMENTS

Under the general heading of carrier modeling we have described, examined, and characterized the carriers within a semiconductor under "rest" or equilibrium conditions. The many important topics addressed in this chapter included the introduction of two "visualization" models; namely, the bonding model and the energy band model. The extremely useful energy band model is actually more than just a model — it is a sophisticated sign language providing a concise means of communicating on a nonverbal level. Relative to the carriers themselves, the reader by now has been successfully prodded into thinking of electrons and holes as classical ball-like "particles," where the charge on an electron is $-q$, the charge on a hole is $+q$, and the effective masses of the particles are m_n^* and m_p^*, respectively. The reader should also know that the carrier numbers in an intrinsic material are equal and relatively small; the carrier concentrations, however, can be selectively increased by adding special impurity atoms or dopants to the semiconductor.

In addressing the problem of determining the carrier concentrations in doped semiconductors, we developed or derived a large number of useful mathematical relationships. The density of states functions [Eqs. (2.6)], the Fermi function [Eq. (2.7)], the symmetrical nondegenerate relationships for n and p [Eqs. (2.19)], the np product [Eq. (2.22)], the charge neutrality relationship [Eq. (2.25)], and the simplified n and p expressions appropriate for a typical semiconductor maintained at room temperature [Eqs. (2.33) and (2.34)] deserve special mention. With regard to using the cited relationships, the reader should be cautioned against "no-think plug and chug." Because semiconductor problems are replete with exceptions, special cases, and nonideal situations, it is imperative that the formula user be aware of derivational assumptions and the validity limits of any and all expressions used in an analysis or computation. In addition to the quantitative carrier relationships, the reader should also have a qualitative "feel" for the carrier distributions in the respective energy bands, the temperature dependence of the intrinsic carrier concentration, and the typical temperature dependence of the majority carrier concentration in a doped semiconductor.

Finally, some attention should be given to the many technical terms and the key parametric values presented in this chapter. The terms extrinsic semiconductor, donor, acceptor, nondegenerate semiconductor, Fermi level, etc., will be encountered again and again in the discussion of semiconductor devices. Likewise, a knowledge of typical numerical values for key parameters, such as $E_G = 1.12$ eV and $n_i \cong 10^{10}/cm^3$ in Si at room temperature, will be invaluable in subsequent work when performing crude, order-of-magnitude computations. Key parametric values also serve as "yardsticks" for gauging whether newly encountered quantities are relatively small or relatively large.

PROBLEMS

2.1 Using the *bonding* model for a semiconductor, indicate how one visualizes:

(a) a donor   (b) an acceptor   (c) a missing atom   (d) a hole

2.2 Using the *energy band* model for a semiconductor, indicate how one visualizes:

(a) an electron   (b) a hole   (c) donor sites   (d) acceptor sites

2.3 Indicate what information is conveyed by the diagrams in Fig. P2.3.

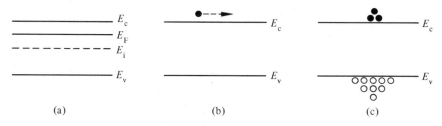

(a)   (b)   (c)

Figure P2.3

2.4 Suppose m_n^* in Si was made smaller somehow. What effect would the reduction in m_n^* have on the density of states in the conduction band, N_C, n_i, and the position of E_i at room temperature?

2.5 Compute the approximate values of N_C and N_V in Si at room temperature using m^* data available in the text. $m_0 = 9.11 \times 10^{-31}$kg; $h = 6.624 \times 10^{-34}$ joul-sec; and $q = 1.6 \times 10^{-19}$coul.

2.6 For a particular semiconductor sample the probability of finding electrons in states an energy kT above the bottom of the conduction band is e^{-11} at room temperature. Determine the location of the Fermi level with respect to E_c in the given material.

2.7 Using the same energy scale (that is, avoiding the use of kT units as shown in Fig. 2.15(b)), plot the Fermi function versus energy E from $E = E_F - 0.15$ eV to $E = E_F + 0.15$ eV for $T = 200°$K, $T = 300°$K and $T = 400°$K. What is the point of this exercise?

2.8 Starting from Eq. (2.8b) and following a procedure analogous to that outlined in the text, present the intermediate steps and arguments leading to Eqs. (2.14b) and (2.16b).

2.9 Suppose that the hole concentration in a piece of Si maintained at room temperature under equilibrium conditions is $10^6/cm^3$.

(a) What is the electron concentration?

(b) Where is the Fermi level positioned relative to the intrinsic Fermi level in the sample?

(c) Draw the energy band diagram for the given material.

2.10 A Si sample is doped with 6×10^{14}As atoms/cm^3.

(a) What is the electron concentration in the Si at room temperature?

(b) What is the electron concentration in the same sample at 500°K?

2.11 A Si sample maintained at room temperature is uniformly doped with $N_D = 10^{14}/cm^3$ donors *and* $N_A = 10^{14}/cm^3$ acceptors.

(a) What are the equilibrium n and p values inside the semiconductor? *Hint:* Examine the charge neutrality relationship.

(b) What happens to the electrons lost by the donor sites? Answer this question using the energy band model with both donor and acceptor levels added to the diagram.

(The canceling action of the two types of dopants as described in this problem is referred to as "compensation.")

2.12 According to the text, the maximum nondegenerate donor doping concentration in Si at room temperature is $N_D \simeq 1.6 \times 10^{18}/cm^3$. Verify the text statement. (An answer within $\pm 10\%$ of the stated value is acceptable.)

2.13 Given an $N_D = 10^{14}/cm^3$ doped Si sample,

(a) Present a *qualitative* argument which leads to the approximate positioning of the Fermi level in the material as $T \rightarrow 0°K$.

(b) Compute E_F as a function of T in the material at 50°K intervals from $T = 300°K$ to $T = 500°K$.

(c) What do you conclude relative to the general behavior of the Fermi level positioning as a function of temperature?

3 / Carrier Action

Carrier modeling under "rest" or equilibrium conditions, considered in the previous chapter, is important because it establishes the proper frame of reference. From a device standpoint, however, the zero current observed under equilibrium conditions is singularly uninteresting. It is only when a semiconductor system is perturbed, giving rise to carrier action or a net carrier response, that currents can flow within and external to the semiconductor system. Action, carrier action, is the general concern of this chapter.

Under normal operating conditions the three primary types of carrier action occurring inside semiconductors are *drift*, *diffusion*, and *recombination–generation*. In this chapter we first describe each primary type of carrier action qualitatively and then quantitatively relate the action to the current flowing within the semiconductor. Special emphasis is placed on characterizing the "constants of the motion" associated with each type of action and, wherever appropriate, the discussion is extended to subsidiary topics of a relevant nature. Although introduced individually, the various types of carrier action are understood to occur simultaneously inside any given semiconductor. Putting everything together, so to speak, next leads to the culmination of our carrier-action efforts; we obtain the basic set of starting equations employed in solving device problems of an electrical nature. Finally, the chapter concludes with an examination of sample problems and problem solutions.

3.1 DRIFT

3.1.1 Definition–Visualization

Drift, by definition, *is charged-particle motion in response to an applied electric field*. Within semiconductors the drifting motion of the carriers on a microscopic scale can be described as follows: When an electric field ($\mathscr{E}$) is applied across a semiconductor as visualized in Fig. 3.1(a), the resulting force on the carriers tends to accelerate the $+q$ charged holes in the direction of the electric field and the $-q$ charged electrons in the direction opposite to the electric field. Because of collisions with ionized impurity atoms and thermally agitated lattice atoms, however, the carrier acceleration is

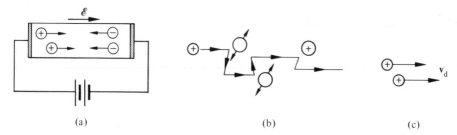

Fig. 3.1 Visualization of carrier drift: (a) motion of carriers within a biased semiconductor bar; (b) drifting hole on a microscopic or atomic scale; (c) carrier drift on a macroscopic scale.

frequently interrupted (the carriers are said to be scattered). The net result, pictured in Fig. 3.1(b), is carrier motion generally along the direction of the electric field, but in a disjointed fashion involving repeated periods of acceleration and subsequent decelerating collisions.

The microscopic drifting motion of a single carrier is obviously complex and quite difficult to analyze quantitatively in any detail. Fortunately, measurable quantities are *macroscopic* observables which reflect the average or overall motion of the carriers. Averaging over all electrons or holes at any given time, we find that the resultant motion of each carrier type can be described in terms of a constant drift velocity, v_d. In other words, on a macroscopic scale, drift may be visualized (see Fig. 3.1(c)) as nothing more than all carriers of a given type moving along at a constant velocity in a direction parallel or antiparallel to the applied electric field.

By way of clarification, it is important to point out that the drifting motion of the carriers arising in response to an applied electric field is actually superimposed upon the always-present thermal motion of the carriers. Electrons in the conduction band and holes in the valence band gain and lose energy via collisions with the semiconductor lattice and are nowhere near stationary even under equilibrium conditions. In fact, under equilibrium conditions the thermally related carrier velocities average ~1/1000 the speed of light at room temperature! As pictured in Fig. 3.2, however, the thermal motion of the carriers is completely random. Thermal motion therefore averages out to zero on a macroscopic scale, does not contribute to current transport, and can be conceptually neglected.

3.1.2 Drift Current

Let us next turn to the task of determining the current flowing within a semiconductor as a result of carrier drift. By definition

I (current) = the charge per unit time crossing an arbitrarily chosen plane of observation oriented normal to the direction of current flow.

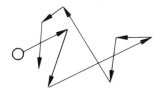

Fig. 3.2 Thermal motion of a carrier.

Considering the p-type semiconductor bar of cross-sectional area A shown in Fig. 3.3, and specifically noting the arbitrarily chosen $\mathbf{v}_d$-normal plane lying within the bar, we can argue:

$v_d t$	All holes this distance back from the $\mathbf{v}_d$-normal plane will cross the plane in a time t;
$v_d t A$	All holes in this volume will cross the plane in a time t;
$p v_d t A$	Holes crossing the plane in a time t;
$q p v_d t A$	Charge crossing the plane in a time t;
$q p v_d A$	Charge crossing the plane per unit time.

The word definition of the last quantity is clearly identical to the formal definition of current. Thus

$$I_{P|\text{drift}} = q p v_d A \qquad \text{hole drift current} \qquad (3.1)$$

As a practical matter, the cross-sectional area A appearing in Eq. (3.1) and other current formulas is often excess baggage. Current, moreover, is generally thought of as a scalar quantity, while in reality it is obviously a vector. These deficiencies are overcome by introducing a related parameter known as the current density, $\mathbf{J}$. $\mathbf{J}$ has the same orientation as the direction of current flow and is equal in magnitude to the current per unit area (or $J = I/A$). By inspection, the current density associated with hole drift

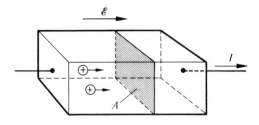

Fig. 3.3 Expanded view of a biased p-type semiconductor bar of cross-sectional area A.

is simply,

$$\mathbf{J}_{P|drift} = qp\,\mathbf{v}_d \tag{3.2}$$

Since the drift current arises in response to an applied electric field, it is reasonable to proceed one step further and seek a form of the current relationship, a modified form of Eq. (3.2), which explicitly relates $\mathbf{J}_{drift}$ to the applied electric field. The desired relationship is readily obtained by making use of an empirical fact; namely, examining experimentally derived v_d versus $\mathscr{E}$ plots of the type displayed in Fig. 3.4, one finds that the drift velocity is a linear function of the applied electric field, $v_d \propto \mathscr{E}$, except for large $\mathscr{E}$-fields encountered in devices only within certain regions or under special conditions. Excluding situations involving large $\mathscr{E}$-fields we can therefore write

$$\mathbf{v}_d = \mu_p \mathscr{E} \tag{3.3}$$

where μ_p, the hole mobility, is the constant of proportionality between v_d and $\mathscr{E}$. Hence, substituting Eq. (3.3) into Eq. (3.2), one obtains

$$\left[\mathbf{J}_{P|drift} = q\mu_p p\mathscr{E}\right] \tag{3.4a}$$

and by a similar argument for electrons,

$$\left[\mathbf{J}_{N|drift} = q\mu_n n\mathscr{E}\right] \tag{3.4b}$$

where μ_n is the electron mobility.*

3.1.3 Mobility

Mobility is obviously a central parameter in characterizing electron and hole transport due to drift. As further readings will reveal, the carrier mobilities also play a key role in characterizing the performance of many devices. It is reasonable therefore to examine μ_n and μ_p in some detail to enhance our general familiarity with the parameters and to establish a core of useful information for future reference.

1. Standard units: cm^2/V-sec.
2. Sample numerical values: $\mu_n \cong 1360\ cm^2/V$-sec in $N_D = 10^{14}/cm^3$ doped silicon at room temperature; $\mu_p \cong 490\ cm^2/V$-sec in $N_A = 10^{14}/cm^3$ doped silicon at room temperature. [The cited values are useful for comparison purposes and when performing "guesstimate" computations. Also note that $\mu_n > \mu_p$. In the major semiconductors μ_n is always greater than μ_p for a given doping level and system temperature.]
3. Physical interpretation: Mobility is a measure of the ease of carrier motion within a semiconductor crystal. A low mobility implies the carriers inside the semiconductor are suffering a relatively large number of motion-impeding collisions. A large mobility, on the other hand, implies the carriers are zipping along with comparative ease.

---

*Although electrons drift in the direction opposite to the applied electric field ($\mathbf{v}_d = -\mu_n\mathscr{E}$), the current transported by negatively charged particles is in turn counter to the direction of drift ($\mathbf{J}_N = -qn\mathbf{v}_d$). The net result, as indicated in Eq. (3.4b), is an electron current in the direction of the applied electric field. It is also important to note, relative to Eqs. (3.3) and (3.4), that μ_n and μ_p are defined to be positive-definite constants.

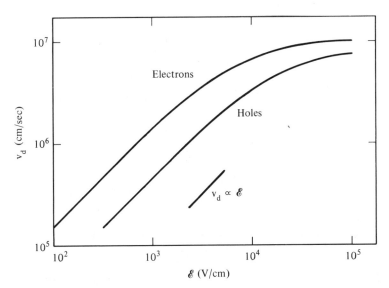

Fig. 3.4 Measured drift velocity of the carriers in ultrapure silicon maintained at room temperature as a function of the applied field.

4. Relationship to scattering: It should be obvious from the preceding (item 3) discussion that carrier mobility varies inversely with the amount of scattering taking place within the semiconductor. Quite simply, an increase in motion-impeding collisions leads to a decrease in mobility. As visualized in Fig. 3.1(b), the dominant scattering mechanisms in nondegenerately doped materials of device quality are typically (i) lattice scattering involving collisions with thermally agitated lattice atoms, and (ii) ionized impurity (i.e., donor-site and/or acceptor-site) scattering.

5. Doping dependence: Figure 3.5 exhibits the observed doping dependence of the electron and hole mobilities in silicon at room temperature. The figure is also typical of the doping dependence observed in other semiconductor materials. At low doping concentrations, below approximately $2 \times 10^{14}/cm^3$ in Si, the carrier mobilities are essentially independent of the doping concentration. For dopings in excess of $\sim 2 \times 10^{14}/cm^3$, the mobilities monotonically decrease with increasing N_A or N_D.

    The explanation of the observed doping dependence is relatively straightforward. At sufficiently low doping levels ionized impurity scattering can be neglected compared to lattice scattering. When lattice scattering, which is not a function of N_A or N_D, becomes the dominant scattering mechanism, it automatically follows that the carrier mobilities will be likewise independent of N_A or N_D. For dopings in excess of $\sim 2 \times 10^{14}/cm^3$, ionized impurity scattering can no longer be neglected. Increasing the number of scattering centers by adding more and more acceptors or donors in excess of $2 \times 10^{14}/cm^3$ clearly increases the amount of nonnegligible ionized impurity scattering and correspondingly decreases the carrier mobilities.

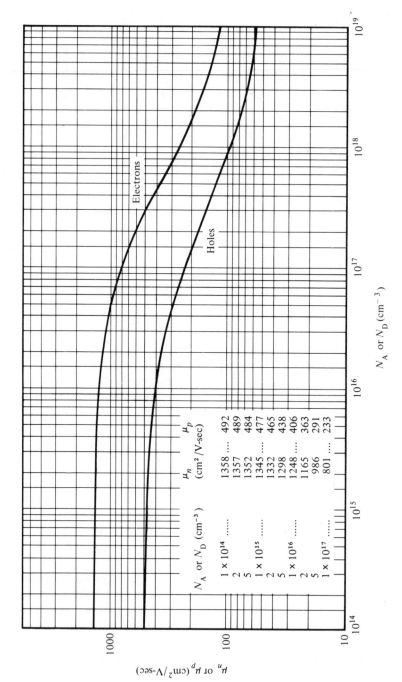

Fig. 3.5 Room temperature carrier mobilities in silicon as a function of the dopant concentration. μ_n is the electron mobility; μ_p is the hole mobility. (Mobility values are based on the empirical relationship suggested by D. M. Caughey and R. F. Thomas, *Proc. IEEE* 55: (1967) 2192. The fit parameters of C. Baccarani and P. Ostoja, *Solid-St. Electron.* 18: (1975) 579, were used in calculating the electron mobility.)

6. Temperature dependence: The temperature dependence of the electron and hole mobilities in low, moderate, and highly doped Si samples is displayed in Fig. 3.6. In the lowest-doped samples the carrier mobilities monotonically decrease as the temperature is increased, where roughly $\mu_n \propto T^{-2.4}$ and $\mu_p \propto T^{-2.2}$. For higher sample dopings the temperature dependence becomes increasingly more complex, with the mobility versus temperature plots clearly exhibiting a mobility maximum in the case of the heavily doped samples.

The general temperature dependence of the carrier mobilities in the lightly doped samples is relatively easy to explain. Increasing the system temperature causes an ever-increasing thermal agitation of the semiconductor atoms, which in turn increases the lattice scattering. Since lattice scattering is the dominant scattering mechanism in the lowest-doped samples, increasing the system temperature therefore monotonically decreases the mobility of the carriers.

For higher sample dopings the increased complexity of the temperature dependence is related to the added effects of ionized impurity scattering. Ionized impurity scattering is not only significant in the more heavily doped samples, but may even become the dominant scattering mechanism at lower temperatures. This is true because, unlike lattice scattering, ionized impurity scattering exhibits an inverse temperature dependence, *decreasing* as the temperature is increased. Although we will omit the details, an appropriate combination of the lattice scattering component, which increases with temperature, and the ionized impurity scattering component, which decreases with temperature, can account for any of the curves presented in Fig. 3.6.

3.1.4 Resistivity

Resistivity is an important material parameter which is closely related to carrier drift. Qualitatively, resistivity is a measure of a material's inherent resistance to current transport — a "normalized" resistance that does not depend on the physical dimensions of the material. Quantitatively, a general expression for the semiconductor resistivity, ρ, can be readily established by again considering the biased semiconductor bar pictured in Fig. 3.1(a). If the bar is assumed to be uniformly doped, only drift currents will be present within the bar under steady state conditions, and one can write

$$\mathbf{J}_{\text{drift}} = \mathbf{J}_{N|\text{drift}} + \mathbf{J}_{P|\text{drift}} = q(\mu_n n + \mu_p p)\mathscr{E} \tag{3.5}$$

However,

$$\mathbf{J}_{\text{drift}} \equiv \sigma \mathscr{E} \equiv \frac{1}{\rho}\mathscr{E} \tag{3.6}$$

In other words, by definition, conductivity (σ) or 1/resistivity is simply the proportionality constant between the drift current density and the applied electric field. Obviously

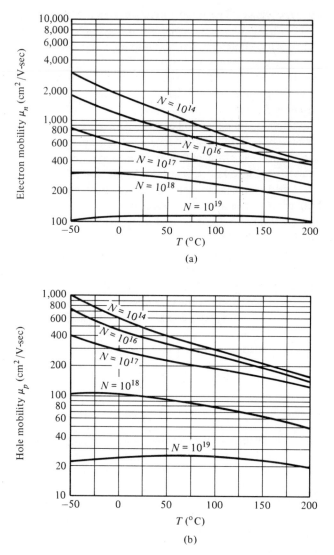

Fig. 3.6 Temperature dependence of (a) electron and (b) hole mobilities in silicon for sample dopings ranging from $10^{14}/cm^3$ to $10^{19}/cm^3$. (From A. B. Phillips, *Transistor Engineering and Introduction to Semiconductor Circuits*. ©1962 by McGraw-Hill Book Co., New York.)

therefore,

$$\left[\rho = \frac{1}{q(\mu_n n + \mu_p p)}\right] \tag{3.7}$$

Having obtained the general expression for ρ, let us next restrict our attention to typical n- and p-type semiconductors maintained at or near room temperature. Specifically, we know from Chapter 2 that $n = N_D$ and $p = n_i^2/N_D \ll n$ in a nondegenerate n-type semiconductor where $N_D \gg n_i$. With $n \gg p$ and $\mu_n \sim \mu_p$ (the electron and hole mobilities in a semiconductor sample typically differ by less than an order of magnitude), we are led to conclude $\mu_n n \gg \mu_p p$. Thus, neglecting $\mu_p p$ in Eq. (3.7) and replacing n by N_D, we obtain

$$\boxed{\rho = \frac{1}{q\mu_n N_D}} \qquad n\text{-type semiconductor} \qquad (3.8a)$$

and similarly,

$$\boxed{\rho = \frac{1}{q\mu_p N_A}} \qquad p\text{-type semiconductor} \qquad (3.8b)$$

It should be noted that, when combined with the mobility versus doping data of Fig. 3.5, Eqs. (3.8) provide a one-to-one correspondence (see Fig. 3.7) between the resistivity, a directly measurable quantity, and the doping inside the semiconductor. As suggested by the foregoing statement, the measured resistivity and appropriate ρ versus doping plot can be used (and, in fact, are routinely used) to determine N_A or N_D.

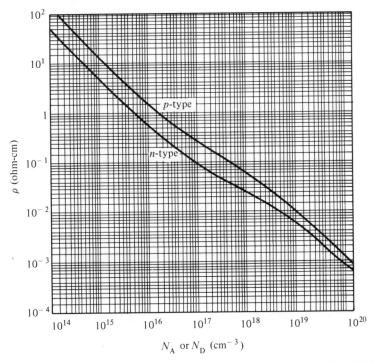

Fig. 3.7 Room temperature resistivity in n- and p-type silicon as a function of the dopant concentration. (From J. C. Irvin, *Bell Syst. Tech. J.* XLI (387) March 1962. ©1962, American Telephone and Telegraph Co. Reprinted by permission.)

The measured resistivity required in determining the doping can be obtained in a number of different ways. A seemingly straightforward approach would be to form the semiconductor into a bar, apply a bias V across contacts attached to the ends of the bar as in Fig. 3.1(a), measure the current I flowing in the circuit, and then deduce ρ from the measured resistance. [R(resistance) $= V/I = \rho l /A$, where l is the bar length and A is the cross-sectional area.] Unfortunately, the straightforward approach is deceptively difficult, is destructive (wastes semiconductor material), and is not readily adaptable to the wafers used in device processing.

The measurement method most widely employed in practice is the four-point probe technique. In the standard four-point probe technique, four collinear, evenly spaced probes, as shown in Fig. 3.8, are brought into contact with the surface of the semiconductor. A known current I is passed through the outer two probes and the potential V thereby developed is measured across the inner two probes. The semiconductor resistivity is then computed from

$$\rho = 2\pi s \frac{V}{I} \mathscr{F} \qquad (3.9)$$

where s is the probe-to-probe spacing and $\mathscr{F}$ is a tabulated or graphed "correction" factor. The correction factor typically depends on the thickness of the sample and on whether the bottom of the semiconductor is touching an insulator or a metal. Unlike the semiconductor-bar measurement, the four-point probe technique is obviously easy to implement, nondestructive, and designed for use in the processing of wafers.*

3.1.5 Band Bending

In previous encounters with the energy band diagram we have consistently drawn E_c and E_v to be energies independent of the position coordinate x. When an electric field giving rise to carrier drift exists inside a semiconductor, the carrier energy versus position dependence is modified and the energy bands are no longer "flat" or position independent; that is, the energy bands "bend" as a function of position.

Seeking to establish the precise relationship between the electric field with a semiconductor and the induced band bending, let us carefully re-examine the energy band diagram. The diagram itself, as emphasized in Fig. 3.9(a), is a plot of the allowed electron energies within the semiconductor as a function of position, where E increasing upward is understood to be the *total* energy of the electrons. Furthermore, we know from previous discussions that if an energy of precisely E_G is added to break a semiconductor-semiconductor bond, the created electron and hole energies would be E_c and E_v, respectively, and the created carriers would be effectively motionless. Absorbing an energy in excess of E_G, on the other hand, would in all probability give rise to an electron energy

---

*For an excellent and detailed discussion of the four-point probe technique, a full tabulation of correction factors, and an examination of other in-use resistivity measurements, the reader is referred to Chapter 3 in W. R. Runyan, *Semiconductor Measurements and Instrumentation*, McGraw-Hill, New York, 1975.

greater than E_c and a hole energy less than E_v, with both carriers moving around rapidly within the lattice. We are led, therefore, to interpret $E - E_c$ to be the kinetic energy (K.E.) of the electrons and $E_v - E$ to be the kinetic energy of the holes (see Fig. 3.9(b)). Moreover, since the total energy equals the sum of the kinetic energy and the potential energy (P.E.), E_c minus the energy reference level (E_{ref}) must equal the electron potential energy, as illustrated in Fig. 3.9(c). [Potential energy, it should be remembered, is arbitrary to within a constant, and hence one can choose any position-independent energy value to be E_{ref}.]

The potential energy is the key in relating the electric field within a semiconductor to positional variations in the energy bands. Specifically, assuming normal operational conditions, where magnetic field, temperature gradient, and stress-induced effects are negligible, only the force associated with an existing electric field can give rise to changes in the potential energy of the carriers. Elementary physics, in fact, tells us that the potential energy of a $-q$ charged particle under such conditions is simply related to the electrostatic potential V at a given point by

$$\text{P.E.} = -qV \tag{3.10}$$

Having previously concluded that

$$\text{P.E.} = E_c - E_{ref} \tag{3.11}$$

we can state

$$\left[V = -\frac{1}{q}(E_c - E_{ref}) \right] \tag{3.12}$$

By definition, moreover,

$$\mathscr{E} = -\nabla V \tag{3.13}$$

or, in one dimension,

$$\mathscr{E} = -\frac{dV}{dx} \tag{3.14}$$

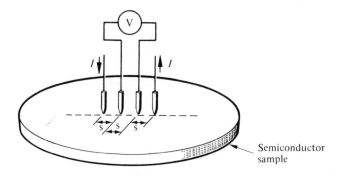

Fig. 3.8 Schematic drawing of the probe arrangement, placement, and biasing in the four-point probe measurement.

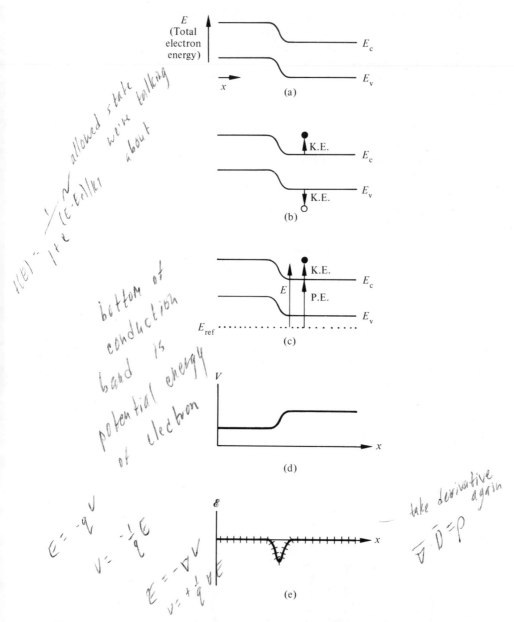

(handwritten annotations surrounding figure):

allowed state talking we're about

$f(E) = \frac{1}{1 + e^{(E-E_f)/kT}}$

bottom of conduction band is potential energy of electron

take derivative again

$\vec{\nabla} \cdot \vec{D} = \rho$

$\mathscr{E} = -\frac{dV}{dx}$

$V = -\frac{1}{q}\mathscr{E}$

$\mathscr{E} = -\nabla V$

$V = +\frac{1}{q} \nabla E$

Fig. 3.9 Relationship between band bending and the electric field within a semiconductor: (a) sample energy band diagram; (b) identification of the carrier kinetic energies; (c) specification of the electron potential energy; (d) electrostatic potential and (e) electric field versus position dependence deduced from and associated with the part (a) energy band diagram.

Consequently,

$$\left[\mathscr{E} = \frac{1}{q}\frac{dE_c}{dx} = \frac{1}{q}\frac{dE_v}{dx} = \frac{1}{q}\frac{dE_i}{dx} \right] \tag{3.15}$$

The latter forms of Eq. (3.15) follow of course from the fact that E_c, E_v and E_i differ by only an additive constant.

The preceding formalization provides a means of readily determining the electric field from the "band bending" in Fig. 3.9(a) and other energy band diagrams. Making use of Eq. (3.12), or essentially just turning E_c in Fig. 3.9(a) upside-down, we obtain the V versus x dependence presented in Fig. 3.9(d). (Please note that V, like P.E., is arbitrary to within a constant; the Fig. 3.9(d) plot can be translated upward or downward along the voltage axis without changing anything physical within the semiconductor.) Finally, taking the slope of E_c versus position, as dictated by Eq. (3.15), produces the $\mathscr{E}$ versus x plot shown in Fig. 3.9(e).

The reader should now be aware of the fact that the energy band diagram also contains information relating to the electrostatic potential and electric field within the semiconductor. Moreover, the general form of the V and $\mathscr{E}$ dependencies within the semiconductor can be obtained almost by inspection. To obtain the general V versus x relationship, merely sketch the upside-down of E_c (or E_v or E_i) versus x; to determine the general $\mathscr{E}$ versus x dependence, simply note the slope of E_c (or E_v or E_i) as a function of position.

3.2 DIFFUSION

3.2.1 Definition – Visualization

Diffusion is a process whereby particles, as a result of their random thermal motion, tend to spread out or redistribute, migrating on a macroscopic scale from regions of high particle concentration toward and into regions of low particle concentration. In the limit, the diffusion process operates so as to produce a uniform distribution of particles. The diffusing entity, it should be noted, need not be charged; thermal motion, not interparticle repulsion, is the enabling action behind the diffusion process.

To cite an everyday, though admittedly somewhat contrived, example of the diffusion process, suppose (after holding your breath) all of the oxygen molecules in the room were gathered together and placed in a corner of the room near the floor. When the O_2 molecules were subsequently released, thermal motion would spread the oxygen throughout the room in a matter of seconds, with intermolecular collisions helping to uniformly redistribute the oxygen to every nook and cranny within the room. This example of diffusion clearly keeps us and the rest of the animal world from periodically gasping for breath.

Seeking to obtain a more detailed understanding of the diffusion process, let us next conceptually "monitor" the process on a microscopic scale employing a simple hypothetical system. The system we propose to monitor is a one-dimensional box containing four compartments and 1024 mobile particles (see Fig. 3.10). The particles within the box obey

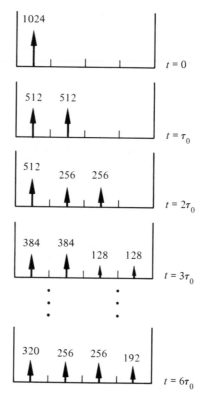

Fig. 3.10 Diffusion on a microscopic scale in a hypothetical one-dimensional system. The numbers over the arrows indicate the number of particles in a given compartment; observation times are listed to the extreme right.

certain stringent rules. Specifically, thermal motion causes all particles in a given compartment to "jump" into an adjacent compartment every τ_0 seconds. In keeping with the random nature of the motion, each and every particle has an equal probability of jumping to the left and to the right. Hitting an "external wall" while attempting to jump to the left or right reflects the particle back to its pre-jump position. Finally, at time $t = 0$ it is assumed that all of the particles are confined in the left-most compartment.

Figure 3.10 records the evolution of our 1024 particle system as a function of time. At time $t = \tau_0$, 512 of the 1024 particles originally in compartment 1 jump to the right and come to rest in compartment 2. The remaining 512 particles jump to the left and are reflected back into the left-most compartment. The end result is 512 particles each in compartments 1 and 2 after τ_0 seconds. At time $t = 2\tau_0$, 256 of the particles in compartment 2 jump into compartment 3, while the remainder jump back into compartment 1. In the meantime, 256 of the particles from compartment 1 jump into compartment 2 and 256 undergo a reflection at the left-hand wall. The net result after $2\tau_0$ seconds is 512 particles in compartment 1 and 256 particles each in compartments 2 and 3. The state of the system after $3\tau_0$ and $6\tau_0$ seconds, also shown in Fig. 3.10, can be deduced in a similar manner. By $t = 6\tau_0$, the particles, once confined to the left-most compartment, have become almost uniformly distributed throughout the box, and it is unnecessary to consider later states. Indeed, the fundamental nature of the diffusion process is clearly self-evident from an examination of the existing states.

In semiconductors the diffusion process on a microscopic scale is similar to that occurring in the hypothetical system except, of course, the random motion of the diffusing particles, the electrons and holes, is three-dimensional and not "compartmentalized." On a macroscopic scale the net effect of the diffusion process is precisely the same within both the hypothetical system and semiconductors. Carrier diffusion on a macroscopic scale, the migration of carriers from regions of high carrier concentration toward and into regions of low carrier concentration, is pictured, for future reference, in Fig. 3.11.

3.2.2 Hot-Point Probe Measurement

Before establishing the current associated with diffusion, let us digress somewhat and briefly consider the hot-point probe measurement. The hot-point probe measurement is a common technique for rapidly determining whether a semiconductor is n- or p-type. From a practical standpoint, knowledge of the semiconductor type is essential in device processing and must be known even before determining the doping concentration from resistivity measurements (refer back to Fig. 3.7). The hot-point probe "typing" experiment is considered here because it simultaneously provides an informative example of the diffusion process.

Examining Fig. 3.12(a), we find the only equipment required in performing the hot-point probe measurement is a hot-point probe, a cold-point probe, and a center-zero milliammeter. The hot-point probe is often a refugee from a wood burning set or simply a soldering iron; the cold-point probe typically assumes the form of an electrical probe like that used with hand-held multimeters. No special requirements are imposed on the center-zero milliammeter connected between the probes. The measurement procedure itself is also extremely simple: After allowing the hot probe to heat up, one brings the two probes into contact with the semiconductor sample, and the ammeter deflects to the right or left, thereby indicating the semiconductor type. Observe that the spacing between the probes is set arbitrarily and can be reduced to enhance the ammeter deflection.

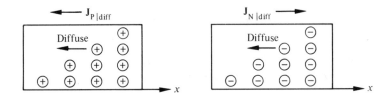

Fig. 3.11 Visualization of electron and hole diffusion on a macroscopic scale.

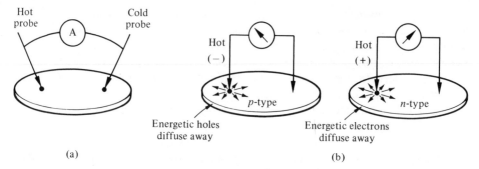

Fig. 3.12   The hot-point probe measurement: (a) required equipment; (b) simplified explanation of how the measurement works.

A simplified explanation of how the hot-point probe measurement works is presented in Fig. 3.12(b). In the vicinity of the probe contact the heated probe creates an increased number of higher energy carriers. These energetic carriers will be predominantly holes in the case of a p-type material and electrons in an n-type material. With more energetic carriers near the heated probe than elsewhere, diffusion acts so as to spread the higher energy carriers throughout the semiconductor wafer. The net effect is a deficit of holes or a net negative charge in the vicinity of the hot-point probe for a p-type material, and a positive charge buildup near the heated probe within an n-type material. Accordingly, the center-zero meter deflects in a different direction for p- and n-type materials.

3.2.3   Diffusion Currents/Total Carrier Currents

In defining the diffusion process and in citing diffusion examples we have sought to emphasize the direct correlation between diffusion and a spatial variation in particle numbers. In order for diffusion to occur, more of the diffusing particles must exist at one point than at other points or, in mathematical terms, there must be a nonzero concentration gradient ($\nabla p \neq 0$ for holes, $\nabla n \neq 0$ for electrons). Logically, moreover, the greater the concentration gradient, the larger the expected flux of carriers and the larger the associated carrier currents. Analysis of the diffusion process indeed confirms the preceding expectations. Currents arising from diffusion are, in fact, directly proportional to the particle concentration gradients. Explicitly, for electrons and holes,

$$\boxed{\mathbf{J}_{P|diff} = -qD_P\nabla p} \tag{3.16a}$$

$$\boxed{\mathbf{J}_{N|diff} = qD_N\nabla n} \tag{3.16b}$$

where the constants of proportionality, D_P and D_N, have units of cm^2/sec and are referred to as the hole and electron diffusion constants, respectively.

Upon examining Eqs. (3.16), please note that the current directions deduced from the equations are consistent with the Fig. 3.11 visualization of the macroscopic diffusion currents. For the positive concentration gradient shown in Fig. 3.11 ($dp/dx > 0$ and $dn/dx > 0$ for the pictured one-dimensional situation), both holes and electrons will diffuse in the $-x$ direction. $J_{P|diff}$ will therefore be negative or directed in the $-x$ direction while $J_{N|diff}$ will be oriented in the $+x$ direction, in complete agreement with Eqs. (3.16).

With explicit expressions available for both the drift and diffusion components of the carrier currents flowing inside of a semiconductor, it is now a trivial matter to establish expressions for the total carrier currents themselves. Simply combining the respective n and p segments of Eqs. (3.4) and Eqs. (3.16), one obtains

$$\mathbf{J_P} = q\mu_p p \mathscr{E} - qD_P \nabla p \qquad (3.17a)$$
$$\updownarrow \text{drift} \qquad \updownarrow \text{diffusion}$$
$$\mathbf{J_N} = q\mu_n n \mathscr{E} + qD_N \nabla n \qquad (3.17b)$$

The double bracketing of Eqs. (3.17) emphasizes the importance of the total-current relationships, which are used directly or indirectly in essentially all device analyses.

3.2.4   Einstein Relationship

Whereas an entire subsection was devoted to examining relevant properties of the carrier mobilities, the constants of the motion associated with drift, we have said little about the diffusion constants, the constants of the motion associated with diffusion. In fact, very little information and even fewer plots of the diffusion constants versus doping and temperature are to be found anywhere in the device literature. What is the reason for this disparity in emphasis? The lack of recorded details about the diffusion constants stems from the existence of a simple relationship interconnecting the diffusion constants and the carrier mobilities. Using the "Einstein relationship," one can readily compute a desired diffusion constant from the corresponding carrier mobility, and to include detailed information about both the diffusion constants and the carrier mobilities would be redundant. In this subsection we present the standard arguments yielding the Einstein relationship.

Let us begin by considering a nondegenerate, *nonuniformly* doped, semiconductor maintained under *equilibrium conditions*. A concrete example of what we have in mind is shown in Fig. 3.13. Figure 3.13(a) exhibits the assumed doping variation with position and Fig. 3.13(b) the corresponding equilibrium energy band diagram. Two facts inherent in the diagram are absolutely essential in establishing the Einstein relationship. First of all, it is a fundamental law of physics that

> under equilibrium conditions the Fermi level inside a material (or inside a group of materials in intimate contact) is invariant as a function of position; that is, $dE_F/dx = dE_F/dy = dE_F/dz = 0$ under equilibrium conditions.

As shown in Fig. 3.13(b), E_F is at the same energy for all x. In a sense the Fermi energy is analogous to the water level in a swimming pool or any other container. Regardless of the contour assumed by the bottom of the pool or container, the water line is always level if the system is unperturbed. Reiterating, $dE_F/dx = 0$ under equilibrium conditions. Secondly, in Chapter 2 we found that the Fermi level in uniformly doped n-type materials moved closer and closer to E_c when the donor doping was systematically increased. Consistent with this fact and as diagrammed in Fig. 3.13(b), the distance between E_c and E_F is smaller in regions of higher doping. Band bending is therefore a natural consequence of the spatial variation in doping, and a nonzero electric field must exist inside non-uniformly doped semiconductors under equilibrium conditions.

Having laid the required foundation, we can now proceed to the derivation proper. Still considering a nondegenerate, nonuniformly doped semiconductor under equilibrium conditions, and simplifying the presentation by working in only one dimension, we can state

$$\mathscr{E} = \frac{1}{q} \frac{dE_i}{dx} \tag{3.18}$$

and

$$n = n_i e^{(E_F - E_i)/kT} \tag{3.19}$$

Moreover, with $dE_F/dx = 0$ under equilibrium conditions,

$$\frac{dn}{dx} = -\frac{n_i}{kT} e^{(E_F - E_i)/kT} \frac{dE_i}{dx} = -\frac{q}{kT} n\mathscr{E} \tag{3.20}$$

We also know that under equilibrium conditions no currents are flowing inside the semiconductor. Thus

$$J_N = q\mu_n n\mathscr{E} + qD_N \frac{dn}{dx} = 0 \tag{3.21}$$

Substituting dn/dx from Eq. (3.20) into Eq. (3.21), and rearranging the result slightly, one obtains

$$(qn\mathscr{E})\mu_n - (qn\mathscr{E})\frac{q}{kT} D_N = 0 \tag{3.22}$$

Since $\mathscr{E} \neq 0$, we therefore conclude

$$\boxed{\frac{D_N}{\mu_n} = \frac{kT}{q}} \qquad \text{Einstein relationship for electrons} \tag{3.23a}$$

A similar argument for holes yields

$$\boxed{\frac{D_P}{\mu_p} = \frac{kT}{q}} \qquad \text{Einstein relationship for holes} \tag{3.23b}$$

Although it was established while assuming equilibrium conditions, we can present more elaborate arguments which show the Einstein relationship to be valid even under nonequilibrium conditions. The nondegenerate restriction, however, still applies; modi-

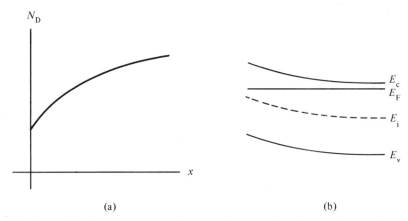

Fig. 3.13 Nonuniformly doped semiconductor: (a) assumed doping variation with position; (b) corresponding equilibrium energy band diagram.

fied, more complex, forms of Eqs. (3.23) result when the argument is extended to degenerate materials. We should also comment that the intermediate results concerning the positional invariance of the equilibrium Fermi level, and the existence of a nonzero electric field inside a nonuniformly doped semiconductor under equilibrium conditions, are important in themselves. Relative to numerical values, it is worthwhile noting that kT/q is a voltage, and at room temperature is approximately equal to 0.026 V.* Hence, for a $N_D = 10^{14}/\text{cm}^3$ Si sample maintained at room temperature, $D_N = (kT/q)\mu_n = (0.026 \text{ V})(1360 \text{ cm}^2/\text{V-sec}) \cong 35 \text{ cm}^2/\text{sec}$. Finally, the Einstein relationship is one of the easiest equations to remember because it rhymes internally — Dee over mu equals kTee over q. The rhyme holds even if the equation is inverted — mu over Dee equals q over kTee.

3.3 RECOMBINATION – GENERATION

3.3.1 Definition – Visualization

Recombination – generation is perhaps the most interesting of the three primary types of carrier action and at the same time the most challenging to totally master. Formally, recombination – generation can be defined as follows:

Generation is a process whereby electrons and holes (carriers) are created.

Recombination is a process whereby electrons and holes are annihilated or destroyed.

---

*At room temperature $kT \cong 0.026$ eV = $(1.6 \times 10^{-19})(0.026)$ joule. Thus $kT/q = (1.6 \times 10^{-19})(0.026)$ $/(1.6 \times 10^{-19})$ joule/coul = 0.026 V.

Unlike drift and diffusion, generation and the inverse action of recombination are not single processes, but instead are collective names for a group of similar processes; that is, there are a number of different ways in which carriers can be created and destroyed within a semiconductor. The simplest of the R – G processes are visualized in Fig. 3.14. Figure 3.14(a) models the photogeneration process where photons (light) with an energy greater than E_G impinge upon the semiconductor and create electrons and holes by breaking semiconductor – semiconductor bonds. Alternatively, the photogeneration process may be visualized in terms of the energy band diagram as light energy being absorbed and exciting electrons from the valence band into the conduction band. The process shown in Fig. 3.14(b) is essentially identical to the Fig. 3.14(a) process, except the excitation is due to thermal (or heat) energy present within the material. Note that the Fig. 3.14(b) action is referred to as *direct* thermal generation. Finally, Fig. 3.14(c) pictures *direct* thermal recombination, the exact inverse of the Fig. 3.14(b) process. An electron and hole moving in the semiconductor lattice, and generally minding their own business, stray into the same spatial vicinity and zap! — the electron and hole annihilate each other.

If a semiconductor happens to be illuminated, photogeneration will typically occur precisely as pictured in Fig. 3.14(a). It is a fact of life, however, that although direct

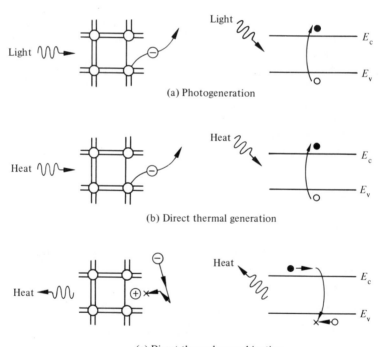

(a) Photogeneration

(b) Direct thermal generation

(c) Direct thermal recombination

Fig. 3.14 Simple recombination–generation processes. Bonding model (left-hand side) and equivalent energy band model (right-hand side) visualization of (a) photogeneration, (b) direct thermal generation, and (c) direct thermal recombination.

thermal generation and direct thermal recombination may be simple to visualize and easy to characterize, the Fig. 3.14(b) and (c) processes are *not* the usual means whereby carriers are thermally created and annihilated in the major semiconductors. In fact, *the thermal creation and annihilation of carriers*, an action going on at all times in all semiconductors, *is typically dominated by indirect thermal recombination – generation*.

Indirect thermal recombination – generation pictured in Fig. 3.15 involves a "third party" or intermediary and takes place only at special locations within the semiconductor known as R – G centers. Physically, R – G centers are special impurity atoms (such as gold, iron, and copper in Si) or lattice defects (such as the missing atom in Fig. 3.15(a)). Lattice defects and the special atoms in the form of unintentional impurities are present even in the major semiconductors, although the R – G center concentration in device-quality materials is normally small compared to the acceptor and donor concentrations. The most important property of the R – G centers is depicted in Fig. 3.15(b); namely, R – G centers (or traps,* for short) introduce allowed electronic levels near the center of the band gap. The near midgap positioning of the E_T levels is all-important because it distinguishes R – G centers from donors and acceptors.

Having provided pertinent R – G center details, let us next briefly examine how carriers are created and destroyed via interacting with the centers. In Fig. 3.15(a) we envision an electron wandering through the semiconductor lattice until it strays into the vicinity of a lattice defect and is caught in the potential well associated with the defect. A little later in time a hole comes along, is attracted to the negatively charged electron, and proceeds to annihilate the electron (and itself) at the R – G center site. The energy band version of precisely the same process is shown on the left-hand side of Fig. 3.15(b). An electron travels to the vicinity of a R – G center, loses energy, and is trapped. Subsequently, a hole wanders into the vicinity of the same center, loses energy, and annihilates

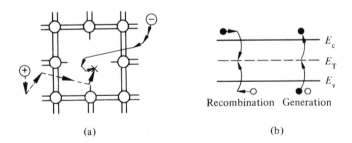

(a)   (b)

Fig. 3.15   Indirect thermal recombination–generation. (a) Bonding model visualization of recombination at a missing atom R–G center. (b) Energy band model description of the R–G center assisted recombination and generation of carriers. R–G centers introduce the E_T levels pictured near the center of the band gap.

---

*To the device specialist the terms R – G center and trap refer to the same type of physical center but describe somewhat different modes of operation. The subscript T, nevertheless, is almost universally employed to identify R – G center parameters such as the E_T energy level in Fig. 3.15(b).

the electron within the center. (Equivalently, one can think of the electron as jumping down a second time and annihilating the hole in the valence band. We should also note it is possible for the hole to be captured first.) The opposite process, indirect thermal generation, is shown on the right-hand side of Fig. 3.15(b). After absorbing thermal energy, an electron first jumps up to the E_T level, creating a hole in the valence band. The creation process is completed when the electron absorbs additional thermal energy and jumps into the conduction band (hence becoming a carrier).

Having completed the examination of basic R–G models and information, we are finally in a position to address the inevitable question of how recombination–generation fits into the overall scheme of things. First of all, it must be understood that the thermal recombination and generation of carriers are never-ending processes. Under equilibrium conditions the recombination and generation rates balance and there is no net change in the carrier numbers. On the other hand, when the semiconductor is perturbed, the thermal recombination and generation rates automatically change so as to favor a return to equilibrium conditions. If the perturbation leads to a carrier excess, the thermal recombination rate becomes greater than the generation rate, thereby favoring a reduction in carrier numbers. If the perturbation causes a carrier deficit, generation will occur at a faster rate than recombination. In other words, thermal R–G is nature's mechanism for stabilizing the carrier concentrations within a material and for reinstating the "most preferred" condition of equilibrium. Photogeneration, we should caution, is a separate process and occurs in addition to thermal R–G when the semiconductor is illuminated. Photogeneration always acts so as to create an excess of carriers. From the foregoing statements, then, we see that the major effect of the R–G action is to produce a change in the carrier concentrations. Recombination–generation indirectly affects current flow in semiconductors by manipulating the carrier concentrations involved in the drift and diffusion processes.

3.3.2   R–G Statistics

R–G statistics is the technical name given to the mathematical characterization of recombination–generation processes. "Mathematical characterization" in the case of recombination–generation does not mean the development of a current density relationship; the R–G action does not lead per se to current transport. Instead, as was pointed out in the last subsection, recombination–generation changes the carrier concentrations, and it is the time rate of change in the carrier concentrations ($\partial n/\partial t$, $\partial p/\partial t$) that must be specified to achieve a mathematical characterization of any R–G process.

To illustrate the general formalism, let us initially consider the simple case of photogeneration. The photogeneration process gives rise to an equal number of added holes and electrons per second, and thus one can write

$$\left.\frac{\partial n}{\partial t}\right|_{\text{light}} = \left.\frac{\partial p}{\partial t}\right|_{\text{light}} = G_{\text{L}} \qquad (\text{number/cm}^3\text{-sec}) \qquad (3.24)$$

The $|_{\text{light}}$ designation means "due to light" and is necessary because n and p can change via other processes. Greater carrier drift or diffusion into a region than out of a region or vice versa, and other forms of recombination – generation, can also change the carrier concentrations. G_L, it should be noted, will in general be a function of the distance the light has penetrated into the material and the wavelength of the light. To achieve closed-form solutions or for illustrative purposes, however, G_L is often taken to be a constant, and the illumination is then said to be uniform.

Thermal R – G statistics, and in particular the statistics for the typically dominant indirect mechanism, are far more complex than the photogeneration statistics. The general case development is, in fact, beyond the scope of this introductory volume. It is possible, nevertheless, to present plausible arguments leading to thermal R – G rate expressions which are valid under circumstances encountered again and again in practical problems.

Let us begin the special case development by defining:

n_0, p_0	carrier concentrations in the material under analysis when equilibrium conditions prevail.
n, p	carrier concentrations in the material under arbitrary conditions.
$\Delta n = n - n_0$; $\Delta p = p - p_0$	deviations in the carrier concentrations from their equilibrium values. Δn and Δp can be both positive and negative, where a positive deviation corresponds to a carrier excess and a negative deviation corresponds to a carrier deficit.
$\left.\dfrac{\partial n}{\partial t}\right\|_{R-G}$, $\left.\dfrac{\partial p}{\partial t}\right\|_{R-G}$	time rate of change in the carrier concentrations due to *both* the indirect thermal recombination and indirect thermal generation processes.
N_T	number of R – G centers/cm^3.

We next require the semiconductor be (1) decidedly n- or p-type and (2) subject to a perturbation which causes only *low level injection*. In Si the first requirement is seldom violated. The second requirement is just an elegant way of saying the perturbation must be relatively small. To be more precise,

> *low level injection implies*
>
> $\Delta p \ll n_0$, $n \simeq n_0$ in an n-type material;
>
> $\Delta n \ll p_0$, $p \simeq p_0$ in a p-type material.

Consider a specific example of $N_D = 10^{14}/\text{cm}^3$ doped Si at room temperature subject to a perturbation where $\Delta p = \Delta n = 10^9/\text{cm}^3$. For the given material $n_0 \simeq N_D = 10^{14}/\text{cm}^3$ and $p_0 \simeq n_i^2/N_D \simeq 10^6/\text{cm}^3$. Consequently, $n = n_0 + \Delta n \simeq n_0$ and $\Delta p = 10^9/\text{cm}^3 \ll n_0 = 10^{14}/\text{cm}^3$. The situation is clearly one of low level injection. Observe, however, that $\Delta p \gg p_0$. Although the majority carrier concentration remains essentially unperturbed under low level injection, the minority carrier concentration can, and routinely does, increase by many orders of magnitude.

With the stage set, so to speak, consider the situation shown in Fig. 3.16. The semiconductor is clearly n-type and a perturbation has caused a $\Delta p \ll n_0$ excess of holes. Assuming the system is in the midst of relaxing back to the equilibrium state from the pictured perturbed state via the indirect thermal R–G process, what factors would you expect to have the greatest effect on the rate of relaxation (i.e., on $\partial p/\partial t|_{R-G}$)? Well, to get rid of an excess hole, the hole must make the transition from the valence band to an electron-filled R–G center. Logically, the greater the number of filled R–G centers, the greater the probability of a hole annihilating transition and the faster the rate of relaxation. Since essentially all of the R–G center levels are filled with electrons because $E_F > E_T$ in our n-type semiconductor, and since the multitude of electrons rapidly fill any levels that become vacant, the number of filled R–G centers during the relaxation process is $\approx N_T$. Thus. we expect $\partial p/\partial t|_{R-G}$ to be approximately proportional to N_T. Moreover, it also seems logical that the number of hole annihilating transitions should increase almost linearly with the number of excess holes. All other factors being equal, the more holes available for annihilation, the greater the number of holes recombining per second. Consequently, we also expect $\partial p/\partial t|_{R-G}$ to be approximately proportional to Δp. Although other factors might be considered, there are no additional dependencies to be established. Hence, introducing a positive proportionality constant, c_p, and realizing that $\partial p/\partial t|_{R-G}$ decreases (is negative) when $\Delta p > 0$, we conclude

$$\left.\frac{\partial p}{\partial t}\right|_{R-G} = -c_p N_T \Delta p \qquad \text{for holes in an } n\text{-type material} \qquad (3.25a)$$

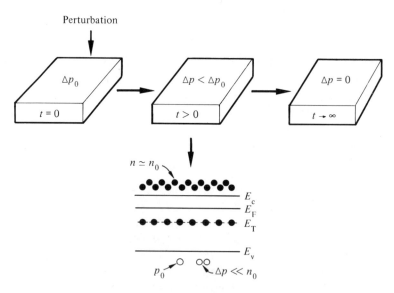

Fig. 3.16 Situation inside an n-type semiconductor after a perturbation causing the low level injection of excess holes.

and by a similar argument,

$$\left.\frac{\partial n}{\partial t}\right|_{\text{R-G}} = -c_n N_T \Delta n \qquad \text{for electrons in a } p\text{-type material} \qquad (3.25\text{b})$$

where c_n is a positive proportionality constant distinct from c_p.

If one examines the left-hand sides of Eqs. (3.25), the dimensional units are found to be those of a concentration divided by time. Since Δp and Δn on the right-hand sides of the same equations are also concentrations, the constants $c_p N_T$ and $c_n N_T$ must have units of 1/time. It is therefore reasonable to introduce the time constants

$$\tau_p = \frac{1}{c_p N_T} \qquad (3.26\text{a})$$

$$\tau_n = \frac{1}{c_n N_T} \qquad (3.26\text{b})$$

which, when substituted into Eqs. (3.25), yield

$$\left[\!\left[\left.\frac{\partial p}{\partial t}\right|_{\text{R-G}} = -\frac{\Delta p}{\tau_p}\right]\!\right] \qquad \text{for holes in an } n\text{-type material} \qquad (3.27\text{a})$$

$$\left[\!\left[\left.\frac{\partial n}{\partial t}\right|_{\text{R-G}} = -\frac{\Delta n}{\tau_n}\right]\!\right] \qquad \text{for electrons in a } p\text{-type material} \qquad (3.27\text{b})$$

Equations (3.27) are the desired end result, the special case characterization of indirect thermal recombination–generation. As indicated previously, the relationships apply only to minority carriers and to situations meeting the low level injection requirement. Technically, steady state or quasisteady state (meaning "nearly" steady state) conditions are also implicitly assumed in the development of Eqs. (3.27). In practice, however, the relationships can be applied with little error to most transient problems of interest. Finally, please note that a $\Delta p < 0$ will give rise to a $\partial p/\partial t|_{\text{R-G}} > 0$. A positive $\partial p/\partial t|_{\text{R-G}}$ simply indicates that a carrier deficit exists inside the semiconductor and generation is occurring at a more rapid rate than recombination. $\partial p/\partial t|_{\text{R-G}}$ and $\partial n/\partial t|_{\text{R-G}}$, it must be remembered, characterize the *net* effect of the thermal recombination and thermal generation processes.

3.3.3 Minority Carrier Lifetimes

The time constants τ_n and τ_p, which were introduced without comment in writing down Eqs. (3.26), are obviously the "constants of the action" associated with recombination–generation. Seeking to provide insight relative to the standard interpretation and naming of the τ's, let us consider once again the situation pictured in Fig. 3.16. Examining the change in the hole concentration with time, it goes almost without saying that the excess holes do not all disappear at the same time. Rather, the hole excess present at $t = 0$ is systematically eliminated, with some of the holes existing for only a short period and others "living" for comparatively long periods of time. If thermal recombination – generation is the sole process acting to relax the semiconductor, the average excess hole lifetime, $\langle t \rangle$, can be computed in a relatively straightforward manner. Without going into details, the computation yields $\langle t \rangle = \tau_n$ (or τ_p). Physically, therefore, τ_n and τ_p have come to be interpreted as *the average time an excess minority carrier will live in a sea of majority carriers*. For identification purposes, τ_n and τ_p are simply referred to as the minority carrier lifetimes.

Given the importance of the τ_n and τ_p parameters in the modeling of devices, one might expect at this point to encounter a sizeable list of relevant facts and information about the minority carrier lifetimes. Actually, a catalog of well-defined properties similar to those of the carrier mobilities just does not exist. In fact, subsidiary experimental measurements must be performed to determine the minority carrier lifetime in a given semiconductor sample. The reason for this apparent lack of information can be traced to the extreme variability of the τ_n and τ_p parameters. Referring to Eqs. (3.26), we see that the carrier lifetimes depend on the often poorly controlled R – G center concentration and *not* on the carefully controlled doping parameters (N_A and N_D). Moreover, the physical nature of the dominant R – G center can vary from sample to sample, and the R – G center concentration changes even within a given sample during device fabrication. A fabrication procedure called "gettering" can reduce the R – G center concentration to a very low level and give rise to a $\tau_n(\tau_p) \sim 1$ msec in Si. The intentional introduction of gold into Si, on the other hand, can controllably increase the R – G center concentration and give rise to a $\tau_n(\tau_p) \sim 1$ nsec. Typical minority carrier lifetimes in most Si devices lie somewhere between the cited extremes; usually $\tau_n(\tau_p)$ are ~ 1 μsec.

3.4 EQUATIONS OF STATE

In the first three sections of this chapter we examined and separately modeled the primary types of carrier action taking place inside semiconductors. Within actual semiconductors the various types of carrier action all occur at the same time, and the state of any semiconductor system can be determined only by taking into account the combined effect of the individual types of carrier action. "Putting it all together," so to speak, leads to the basic set of starting equations employed in solving device problems, herein referred to as the *equations of state*. This section is devoted to developing the equations of state and to surveying common simplifications and special-case solutions.

3.4.1 Continuity Equations

Each and every type of carrier action, whether it be drift, diffusion, thermal recombination, thermal generation, or some other type of carrier action, gives rise to a change in the carrier concentrations with time. The combined effect of all types of carrier action can therefore be taken into account by equating the overall change in the carrier concentrations per unit time ($\partial n/\partial t$ or $\partial p/\partial t$) to the sum of the $\partial n/\partial t$'s or $\partial p/\partial t$'s due to the individual processes; that is,

$$\frac{\partial n}{\partial t} = \frac{\partial n}{\partial t}\bigg|_{\text{drift}} + \frac{\partial n}{\partial t}\bigg|_{\text{diff}} + \frac{\partial n}{\partial t}\bigg|_{\substack{\text{thermal} \\ \text{R}-\text{G}}} + \frac{\partial n}{\partial t}\bigg|_{\substack{\text{other processes} \\ \text{(photogen., etc.)}}} \qquad (3.28a)$$

$$\frac{\partial p}{\partial t} = \frac{\partial p}{\partial t}\bigg|_{\text{drift}} + \frac{\partial p}{\partial t}\bigg|_{\text{diff}} + \frac{\partial p}{\partial t}\bigg|_{\substack{\text{thermal} \\ \text{R}-\text{G}}} + \frac{\partial p}{\partial t}\bigg|_{\substack{\text{other processes} \\ \text{(photogen., etc.)}}} \qquad (3.28b)$$

The overall effect of the individual processes is established, in essence, by invoking the requirement of conservation of carriers. Electrons and holes cannot mysteriously appear and disappear at a given point, but must be transported to or created at the given point via some type of ongoing carrier action. There must be a spatial and time continuity in the carrier concentrations. For this reason Eqs. (3.28) are known as the *continuity equations*.

The continuity equations can be written in a somewhat more compact form by noting

$$\frac{\partial n}{\partial t}\bigg|_{\text{drift}} + \frac{\partial n}{\partial t}\bigg|_{\text{diff}} = \frac{1}{q}\left(\frac{\partial J_{\text{N}x}}{\partial x} + \frac{\partial J_{\text{N}y}}{\partial y} + \frac{\partial J_{\text{N}z}}{\partial z}\right) = \frac{1}{q}\nabla \cdot \mathbf{J}_{\text{N}} \qquad (3.29a)$$

$$\frac{\partial p}{\partial t}\bigg|_{\text{drift}} + \frac{\partial p}{\partial t}\bigg|_{\text{diff}} = -\frac{1}{q}\left(\frac{\partial J_{\text{P}x}}{\partial x} + \frac{\partial J_{\text{P}y}}{\partial y} + \frac{\partial J_{\text{P}z}}{\partial z}\right) = -\frac{1}{q}\nabla \cdot \mathbf{J}_{\text{P}} \qquad (3.29b)$$

Equations (3.29), which can be established by a straightforward mathematical manipulation, merely state that there will be a change in the carrier concentrations within a given section of the semiconductor if more carriers drift and/or diffuse into the region than out of the region (or vice versa). Utilizing Eqs. (3.29), we obtain

$$\boxed{\frac{\partial n}{\partial t} = \frac{1}{q}\nabla \cdot \mathbf{J}_{\text{N}} + \frac{\partial n}{\partial t}\bigg|_{\substack{\text{thermal} \\ \text{R}-\text{G}}} + \frac{\partial n}{\partial t}\bigg|_{\substack{\text{other} \\ \text{processes}}}} \qquad (3.30a)$$

$$\boxed{\frac{\partial p}{\partial t} = -\frac{1}{q}\nabla \cdot \mathbf{J}_{\text{P}} + \frac{\partial p}{\partial t}\bigg|_{\substack{\text{thermal} \\ \text{R}-\text{G}}} + \frac{\partial p}{\partial t}\bigg|_{\substack{\text{other} \\ \text{processes}}}} \qquad (3.30b)$$

The (3.30) continuity equations are completely general and directly or indirectly constitute the starting point in most device analyses. In computer simulations the continuity equations are employed directly. The appropriate relationships for $\partial n/\partial t|_{\text{thermal R}-\text{G}}$, $\partial p/\partial t|_{\text{thermal R}-\text{G}}$ [which may or may not be the special-case relationships cited in Eqs. (3.27)], along with the concentration changes due to "other processes," are substituted into Eqs. (3.30), and numerical solutions are sought for $n(x, y, z, t)$ and $p(x, y, z, t)$. In most problems where a closed-form type of solution is desired, however, the continuity equations are used only in an indirect fashion. The actual starting point in such analyses is a simplified version of the continuity equations to be established in the next subsection.

3.4.2   Minority Carrier Diffusion Equations

The "workhorse" minority carrier diffusion equations are derived from the continuity equations by invoking the following set of simplifying assumptions:

1. The particular system under analysis is *one dimensional*; i.e., all variables are at most a function of just one coordinate (say the x-coordinate).

2. The analysis is limited or restricted to *minority carriers*.

3. $\mathscr{E} \simeq 0$ in the semiconductor or regions of the semiconductor subject to analysis.

4. The equilibrium minority carrier concentrations are not a function of position. In other words, $n_0 \neq n_0(x)$, $p_0 \neq p_0(x)$.

5. *Low level injection* conditions prevail.

6. There are *no* "*other processes*," except possibly photogeneration, taking place within the system.

Working on the continuity equation for electrons, we note that

$$\frac{1}{q} \nabla \cdot \mathbf{J_N} \rightarrow \frac{1}{q} \frac{\partial J_N}{\partial x} \tag{3.31}$$

if the system is one-dimensional (simplification #1). Moreover,

$$J_N = q\mu_n n \mathscr{E} + qD_N \frac{\partial n}{\partial x} \simeq qD_N \frac{\partial n}{\partial x} \tag{3.32}$$

when $\mathscr{E} \simeq 0$ and one is concerned only with minority carriers (simplifications #2 and #3). By way of explanation, the drift component can be neglected in the current density expression because $\mathscr{E}$ is small by assumption and minority carrier concentrations are also small, making the $n\mathscr{E}$ product extremely small. [Note that the same argument cannot be applied to majority carriers.] Since, by assumption (simplification #4), $n_0 \neq n_0(x)$, and by definition $n = n_0 + \Delta n$, we can also write

$$\frac{\partial n}{\partial x} = \frac{\partial n_0}{\partial x} + \frac{\partial \Delta n}{\partial x} = \frac{\partial \Delta n}{\partial x} \tag{3.33}$$

Combining Eqs. (3.31) through (3.33) yields

$$\frac{1}{q} \nabla \cdot \mathbf{J_N} \rightarrow D_N \frac{\partial^2 \Delta n}{\partial x^2} \tag{3.34}$$

Turning to the remaining terms in the continuity equation for electrons, the low level injection restriction (simplification #5) combined with the minority carrier limitation (simplification #2) permits us to employ the special-case expression for $\partial n / \partial t|_{\text{thermal R-G}}$ derived in Section 3.3.

$$\left. \frac{\partial n}{\partial t} \right|_{\substack{\text{thermal} \\ \text{R-G}}} = -\frac{\Delta n}{\tau_n} \tag{3.35}$$

In addition, simplification #6 allows us to write,

$$\left. \frac{\partial n}{\partial t} \right|_{\substack{\text{other} \\ \text{processes}}} = G_L \tag{3.36}$$

where it is understood that $G_L = 0$ if the semiconductor is not subject to illumination.

Finally, the equilibrium electron concentration is never a function of time, $n_0 \neq n_0(t)$, and we can therefore write

$$\frac{\partial n}{\partial t} = \frac{\partial n_0}{\partial t} + \frac{\partial \Delta n}{\partial t} = \frac{\partial \Delta n}{\partial t} \tag{3.37}$$

Substituting Eqs. (3.34) through (3.37) into the (3.30a) continuity equation, and simultaneously recording the analogous result for holes, one obtains

$$\left[\begin{array}{c} \dfrac{\partial \Delta n_\mathrm{p}}{\partial t} = D_\mathrm{N}\dfrac{\partial^2 \Delta n_\mathrm{p}}{\partial x^2} - \dfrac{\Delta n_\mathrm{p}}{\tau_\mathrm{n}} + G_\mathrm{L} \\[2em] \dfrac{\partial \Delta p_\mathrm{n}}{\partial t} = D_\mathrm{P}\dfrac{\partial^2 \Delta p_\mathrm{n}}{\partial x^2} - \dfrac{\Delta p_\mathrm{n}}{\tau_\mathrm{p}} + G_\mathrm{L} \end{array}\right]$$

$$\text{(3.38a)}$$
Minority carrier diffusion equations
$$\text{(3.38b)}$$

We have added subscripts to the carrier concentrations in Eqs. (3.38) to remind the user that the equations are valid only for minority carriers, applying to electrons in p-type materials and to holes in n-type materials.

3.4.3 Simplifications and Solutions

In the course of performing device analyses the conditions of a problem often permit additional simplifications that drastically reduce the complexity of the minority carrier diffusion equations. More often than not, however, a simplification is stated in words, and difficulty is sometimes encountered in perceiving the mathematical consequences of the simplification. To assist the reader, we have tabulated below the most commonly encountered simplification statements and the corresponding mathematical modifications of the minority carrier diffusion equations. We also note below solutions to forms of the minority carrier diffusion equations resulting from certain combinations of the cited simplifications, solutions one is likely to encounter in later readings.

Simplification statement	Mathematical simplification
Steady state	$\dfrac{\partial \Delta n_p}{\partial t} \to 0 \quad \left(\dfrac{\partial \Delta p_n}{\partial t} \to 0\right)$
No concentration gradient or no diffusion current	$D_N \dfrac{\partial^2 \Delta n_p}{\partial x^2} \to 0 \quad \left(D_P \dfrac{\partial^2 \Delta p_n}{\partial x^2} \to 0\right)$
No drift current or $\mathscr{E} = 0$	No further simplification. ($\mathscr{E} \simeq 0$ is assumed in the derivation.)
No thermal R–G	$\dfrac{\Delta n_p}{\tau_n} \to 0 \quad \left(\dfrac{\Delta p_n}{\tau_p} \to 0\right)$
No light	$G_\mathrm{L} \to 0$

Solution no. 1

GIVEN: Steady state, no light.

SIMPLIFIED
DIFF. EQN:
$$0 = D_N \frac{d^2 \Delta n_p}{dx^2} - \frac{\Delta n_p}{\tau_n}$$

SOLUTION: $\Delta n_p(x) = A e^{-x/L_N} + B e^{x/L_N}$
where $L_N \equiv \sqrt{D_N \tau_n}$
and A, B are solution constants.

Solution no. 2

GIVEN: No concentration gradient, no light.

SIMPLIFIED
DIFF. EQN:
$$\frac{d \Delta n_p}{dt} = - \frac{\Delta n_p}{\tau_n}$$

SOLUTION: $\Delta n_p(t) = \Delta n_p(0) e^{-t/\tau_n}$

Solution no. 3

GIVEN: Steady state, no concentration gradient.

SIMPLIFIED
DIFF. EQN:
$$0 = - \frac{\Delta n_p}{\tau_n} + G_L$$

SOLUTION: $\Delta n_p = G_L \tau_n$

Solution no. 4

GIVEN: Steady state, no R$-$G, no light.

SIMPLIFIED
DIFF. EQN:
$$0 = D_N \frac{d^2 \Delta n_p}{dx^2} \quad \text{or} \quad 0 = \frac{d^2 \Delta n_p}{dx^2}$$

SOLUTION: $\Delta n_p(x) = A + Bx$

3.4.4 Equations of State Summary

For the reader's convenience we have collected in this subsection all of the equations routinely encountered in device analyses. Except for (3.41) and (3.44), the equations are a repetition of previously established relationships. The reader is assumed to be familiar with Eq. (3.41) which is a well-known expression from the field of electricity and magnetism. Equation (3.44), on the other hand, merely states that the total current density under steady state conditions (**J**) is simply the sum of the electron and hole current densities. The symbol ρ appearing in Eqs. (3.41) and (3.43), one should note, is the charge density, the charge/cm^3, and not the resistivity. It is unfortunate that the same symbol has come to be closely identified with two different quantities, but this duality never leads to difficulties because the meaning of the symbol can always be inferred from the context of the problem. Also, in writing down the continuity equations we have introduced simplified symbols for the various $\partial n/\partial t$'s and $\partial p/\partial t$'s. The simplified continuity equation symbols and all other newly introduced symbols are defined at the end of the subsection.

$$\frac{\partial n}{\partial t} = \frac{1}{q}\nabla \cdot \mathbf{J}_N - r_N + g_N \qquad \qquad \left(\begin{array}{c}\text{Continuity}\\ \text{equations}\end{array}\right) \qquad (3.39a)$$

$$\frac{\partial p}{\partial t} = -\frac{1}{q}\nabla \cdot \mathbf{J}_P - r_P + g_P \qquad \qquad (3.39b)$$

$$\frac{\partial \Delta n_p}{\partial t} = D_N\frac{\partial^2 \Delta n_p}{\partial x^2} - \frac{\Delta n_p}{\tau_n} + G_L \qquad \qquad (3.40a)$$

$$\frac{\partial \Delta p_n}{\partial t} = D_P\frac{\partial^2 \Delta p_n}{\partial x^2} - \frac{\Delta p_n}{\tau_p} + G_L \qquad \left(\begin{array}{c}\text{Minority carrier}\\ \text{diffusion equations}\end{array}\right) \qquad (3.40b)$$

$$\nabla \cdot \mathscr{E} = \rho/K_S\varepsilon_0 \qquad \qquad \text{(Poisson's equation)} \qquad (3.41)$$

where

$$\mathbf{J}_N = q\mu_n n\mathscr{E} + qD_N\nabla n \qquad \qquad (3.42a)$$

$$\mathbf{J}_P = q\mu_p p\mathscr{E} - qD_P\nabla p \qquad \qquad (3.42b)$$

$$\rho = q(p - n + N_D^+ - N_A^-) \qquad \qquad (3.43)$$

$$\mathbf{J} = \mathbf{J}_N + \mathbf{J}_P \qquad \qquad (3.44)$$

and

$$r_N \equiv -\frac{\partial n}{\partial t}\bigg|_{\text{thermal R–G}} ; \qquad r_P \equiv -\frac{\partial p}{\partial t}\bigg|_{\text{thermal R–G}} ;$$

$$g_N \equiv \frac{\partial n}{\partial t}\bigg|_{\text{other processes}} ; \qquad g_P \equiv \frac{\partial p}{\partial t}\bigg|_{\text{other processes}} ;$$

ρ = charge density, charge/cm^3;

K_S = semiconductor dielectric constant;

ε_0 = permittivity of free space;

$\mathbf{J}$ = the total current density under steady state conditions.

3.5 PROBLEM SOLVING AND SUPPLEMENTAL CONCEPTS

A person encountering the topic of carrier action for the first time is apt to be overwhelmed. How does one use all these formulas? Where does one start in solving actual problems? Device analyses are, of course, examples of problem solutions, and most of the reader's questions will be answered in subsequent discussions. Nevertheless, it is worthwhile to examine some simple problems to indicate suggested problem-solving approaches. More-over, the sample problems reviewed herein serve the additional purpose of introducing two supplemental concepts which are encountered in a number of device analyses; namely, the minority carrier diffusion lengths and quasi-Fermi levels.

3.5.1   Sample Problem No. 1

Problem: A uniformly donor-doped silicon wafer maintained at room temperature is suddenly illuminated with light at time $t = 0$. Assuming $N_D = 10^{15}/\text{cm}^3$, $\tau_p = 10^{-6}\text{sec}$, and a light-induced creation of 10^{17} electrons and holes per cm^3-sec throughout the semiconductor, determine $\Delta p_n(t)$ for $t > 0$.

Solution: *Step 1* — Review precisely what information is given or implied in the statement of the problem.

The semiconductor is silicon, $T = $ room temperature, the donor doping is the same everywhere with $N_D = 10^{15}/\text{cm}^3$, and $G_L = 10^{17}/\text{cm}^3$-sec at all points inside the semiconductor. Also, the statement of the problem *implies* equilibrium conditions exist for $t < 0$.

Step 2 — Characterize the system under equilibrium conditions.

For Si at room temperature $n_i = 10^{10}/\text{cm}^3$. Since $N_D \gg n_i$, $n_0 = N_D = 10^{15}/\text{cm}^3$ and $p_0 = n_i^2/N_D = 10^5/\text{cm}^3$. Because of the uniform doping, the equilibrium n_0 and p_0 values are, of course, the same everywhere throughout the semiconductor.

Step 3 — Analyze the problem qualitatively.

Prior to $t = 0$, equilibrium conditions prevail and $\Delta p_n = 0$. Starting at $t = 0$ the light creates added electrons and holes and Δp_n will begin to increase. The growing excess carrier numbers, however, in turn lead to an enhanced thermal recombination rate, with $\partial p/\partial t|_{R-G} = -\Delta p_n/\tau_p$. Consequently, as Δp_n increases as a result of photogeneration, more and more of the excess holes are eliminated per second by thermal recombination. Eventually, a point is reached where the carriers annihilated per second by thermal recombination balance the carriers created per second by the light, and a steady state condition is attained.

Summarizing, we expect $\Delta p_n(t)$ to start from zero at $t = 0$, build up at a decreasing rate, and ultimately become constant. Since $\partial p/\partial t|_{\text{light}} + \partial p/\partial t|_{R-G} = 0$ or $G_L - \Delta p_n/\tau_p = 0$ under steady state conditions, we can even state that $\Delta p_n(t \to \infty) = G_L\tau_p$.

Step 4 — *Perform a quantitative analysis.*

The minority carrier diffusion equation is the starting point for most first-order quantitative analyses. After examining the problem for obvious conditions which would invalidate the use of the diffusion equation, the appropriate minority carrier diffusion equation is written down, the equation is simplified, and a solution is sought subject to boundary conditions stated or implied in the problem.

For the problem under consideration a cursory inspection reveals that all simplifying assumptions involved in deriving the diffusion equations are readily satisfied. Specifically, only the minority carrier concentration is of interest; the equilibrium carrier concentrations are not a function of position, and there are no "other processes" except for photogeneration. Because the photogeneration is uniform throughout the semiconductor, the perturbed carrier concentrations are also position-independent ($\Delta p_n \neq \Delta p_n(x, y, z)$) and the electric field $\mathscr{E}$ must clearly be zero in the perturbed system. Finally, since $\Delta p_{n|max} = G_L \tau_p = 10^{11}/cm^3 \ll n_0 = 10^{15}/cm^3$, low level injection conditions prevail at all times.

With no obstacles to utilizing the diffusion equation, the desired quantitative solution can now be obtained by solving

$$\frac{\partial \Delta p_n}{\partial t} = D_P \frac{\partial^2 \Delta p_n}{\partial x^2} - \frac{\Delta p_n}{\tau_p} + G_L \qquad (3.45)$$

subject to the boundary condition

$$\Delta p_n(t)|_{t=0} = 0 \qquad (3.46)$$

Since $\Delta p_n \neq \Delta p_n(x)$, the diffusion equation simplifies to

$$\frac{d \Delta p_n}{dt} + \frac{\Delta p_n}{\tau_p} = G_L \qquad (3.47)$$

The general solution of Eq. (3.47) is

$$\Delta p_n(t) = G_L \tau_p + A e^{-t/\tau_p} \qquad (3.48)$$

Applying the boundary condition yields

$$A = -G_L \tau_p \qquad (3.49)$$

and

$$\Delta p_n(t) = G_L \tau_p (1 - e^{-t/\tau_p}) \qquad \Leftarrow solution \qquad (3.50)$$

Step 5 — Examine the solution.

Failing to examine the mathematical solution to a problem is like growing vegetables and then failing to eat the produce. Relative to the Eq. (3.50) result, $G_L \tau_p$ has the dimensions of a concentration (number/cm^3) and the solution is at least dimensionally correct. A plot of the Eq. (3.50) result is shown in Fig. 3.17. Note that, in agreement with qualitative predictions, $\Delta p_n(t)$ starts from zero at $t = 0$ and eventually saturates at $G_L \tau_p$ after a few τ_p.

3.5.2 Sample Problem No. 2

Problem: As pictured in Fig. 3.18(a), the $x = 0$ end of a uniformly doped semi-infinite bar of silicon with $N_D = 10^{15}/cm^3$ is illuminated so as to create $\Delta p_{n0} = 10^{10}/cm^3$ excess holes at $x = 0$. The wavelength of the illumination is such that no light penetrates into the interior ($x > 0$) of the bar. Determine $\Delta p_n(x)$.

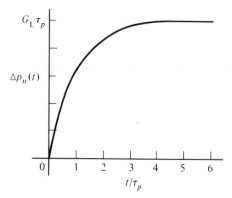

Fig. 3.17 Solution to Sample Problem No. 1. Photogeneration-induced increase in the excess hole concentration as a function of time.

Solution: The semiconductor is again silicon uniformly doped with an $N_D = 10^{15}/\text{cm}^3$. Steady state conditions are inferred from the statement of the problem, since we are asked for $\Delta p_n(x)$ and not $\Delta p_n(x, t)$. Moreover, at $x = 0$ $\Delta p_n(0) = \Delta p_{n0} = 10^{10}/\text{cm}^3$ and $\Delta p_n \to 0$ as $x \to \infty$. The latter boundary condition follows from the semi-infinite nature of the bar. The perturbation in Δp_n due to the nonpenetrating light can't possibly extend out to $x = \infty$. The nonpenetrating nature of the light itself tells us that $G_L = 0$ for $x > 0$. Finally, note that the problem statement fails to mention the temperature of operation. When this happens it is reasonable to assume an intended $T = $ room temperature.

If we take $T = $ room temperature, the silicon bar in Problem 2 would revert to an equilibrium condition identical to that described in Problem 1 if the light were removed.

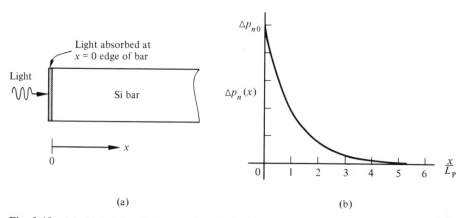

(a)   (b)

Fig. 3.18 (a) Pictorial definition of Sample Problem No. 2. (b) Solution to Sample Problem No. 2 showing the excess hole concentration inside the Si bar as a function of position.

Under equilibrium conditions, then, $n_0 = 10^{15}/\text{cm}^3$, $p_0 = 10^5/\text{cm}^3$, and the carrier concentrations are uniform throughout the semiconductor bar.

Qualitatively it is a simple matter to predict the expected effect of the nonpenetrating light on the silicon bar. The light first creates excess carriers right at $x = 0$. With more carriers at one point than elsewhere in the bar, the diffusion process next comes into play and the carrier excess spreads into the semiconductor proper. At the same time, however, the appearance of an excess hole concentration inside the bar enhances the thermal recombination rate. Thus, as the diffusing holes move into the bar their numbers are reduced by recombination. In addition, since the minority carrier holes live for only a limited time period, a time τ_p on the average, fewer and fewer excess holes will survive as x becomes larger and larger. Under steady state conditions it is therefore reasonable to expect an excess distribution of holes near $x = 0$, with $\Delta p_n(x)$ monotonically decreasing from Δp_{n0} at $x = 0$ to $\Delta p_n = 0$ as $x \rightarrow \infty$.

In preparation for obtaining a quantitative solution we observe that the system under consideration is one-dimensional, the analysis is restricted to the minority carrier holes, the equilibrium carrier concentrations are position-independent, there are no "other processes" for $x > 0$, and low level injection conditions clearly prevail ($\Delta p_{n|\text{max}} = \Delta p_{n0} = 10^{10}/\text{cm}^3 \ll n_0 = 10^{15}/\text{cm}^3$). The only question that might be raised concerning the use of the diffusion equation as the starting point for the quantitative analysis is whether $\mathscr{E} \simeq 0$. With the light on, a nonuniform distribution of holes and associated distribution of positive charge will appear near the $x = 0$ surface. *A priori*, however, it is impossible to ascertain whether the $\mathscr{E} \simeq 0$ assumption is violated; therefore the usual procedure in problems of this sort is to proceed with the analysis using the diffusion equation and to subsequently check the resulting solution for inconsistencies.

Under steady state conditions with $G_L = 0$ for $x > 0$ the hole diffusion equation reduces to the form

$$D_P \frac{d^2 \Delta p_n}{dx^2} - \frac{\Delta p_n}{\tau_p} = 0 \qquad \text{for } x > 0 \tag{3.51}$$

which is to be solved subject to the boundary conditions

$$\Delta p_{n|x=0^+} = \Delta p_{n|x=0} = \Delta p_{n0} \tag{3.52}$$

and

$$\Delta p_{n|x\rightarrow\infty} = 0 \tag{3.53}$$

Equation (3.51) should be recognized as one of the simplified diffusion equations cited in Subsection 3.4.3, with the general solution

$$\Delta p_n(x) = A e^{-x/L_P} + B e^{x/L_P} \tag{3.54}$$

where

$$L_P \equiv \sqrt{D_P \tau_p} \tag{3.55}$$

Because $\exp(x/L_P) \rightarrow \infty$ as $x \rightarrow \infty$, the only way that the Eq. (3.53) boundary condition can be satisfied is for B to be identically zero. With $B = 0$, application of the Eq. (3.52) boundary condition yields

$$A = \Delta p_{n0} \qquad (3.56)$$

and

$$\Delta p_n(x) = \Delta p_{n0} e^{-x/L_P} \qquad \Leftarrow \text{solution} \qquad (3.57)$$

The Eq. (3.57) result is plotted in Fig. 3.18(b). In agreement with qualitative arguments, the nonpenetrating light merely gives rise to a monotonically decreasing $\Delta p_n(x)$ starting from Δp_{n0} at $x = 0$ and decreasing to $\Delta p_n = 0$ as $x \to \infty$. Note that the precise functional form of the falloff in the excess carrier concentration is exponential with a characteristic decay length equal to L_P.

3.5.3 Diffusion Lengths

The situation just encountered in Sample Problem No. 2 — the creation (or appearance) of an excess of minority carriers along a given plane in a semiconductor, the subsequent diffusion of the excess from the point of injection, and the exponential falloff in the excess carrier concentration characterized by a decay length (L_P) — occurs often enough in semiconductor analyses that the characteristic length has been given a special name. Specifically,

$$L_P \equiv \sqrt{D_P \tau_p} \qquad \begin{array}{l} \text{associated with the minority carrier} \\ \text{holes in an } n\text{-type material} \end{array} \qquad (3.58a)$$

and

$$L_N \equiv \sqrt{D_N \tau_n} \qquad \begin{array}{l} \text{associated with the minority carrier} \\ \text{electrons in a } p\text{-type material} \end{array} \qquad (3.58b)$$

are referred to as *minority carrier diffusion lengths*.

Physically, L_P and L_N represent the average distance minority carriers can diffuse into a sea of majority carriers before being annihilated. This interpretation is clearly consistent with Sample Problem No. 2, where the average position of the excess minority carriers inside the semiconductor bar is

$$\langle x \rangle = \int_0^\infty x \Delta p_n(x)\, dx \Big/ \int_0^\infty \Delta p_n(x)\, dx = L_P \qquad (3.59)$$

For memory purposes minority carrier diffusion into a sea of majority carriers might be likened to a group of swimmers attempting to cross a piranha-infested stretch of the Amazon River. In the analogy L_P and L_N correspond to the average distance the swimmers advance into the river before being eaten.

To obtain some idea as to the size of diffusion lengths, let us assume $T = $ room temperature, an $N_D = 10^{15}/\text{cm}^3$ doped semiconductor, and a $\tau_p = 10^{-6}$ sec. For the given system

$$L_P = \sqrt{D_P \tau_p} = \sqrt{(kT/q)\mu_p \tau_p} = [(0.026)(477)(10^{-6})]^{1/2}$$
$$\simeq 3.5 \times 10^{-3} \, cm$$

3.5.4 Quasi-Fermi Levels

Quasi-Fermi levels are energy levels used to specify the carrier concentrations inside a semiconductor under *non*equilibrium conditions.

To understand the need for introducing quasi-Fermi levels let us first refer to Sample Problem No. 1. In this problem equilibrium conditions prevailed prior to $t = 0$, with $n_0 = N_D = 10^{15}/cm^3$ and $p_0 = 10^5/cm^3$. The energy band diagram describing the equilibrium situation is shown in Fig. 3-19(a). A simple inspection of the energy band diagram and the Fermi level positioning also conveys, of course, the equilibrium carrier concentrations, since

$$n_0 = n_i e^{(E_F - E_i)/kT} \tag{3.60a}$$

$$p_0 = n_i e^{(E_i - E_F)/kT} \tag{3.60b}$$

The point we wish to emphasize is that under equilibrium conditions there is a one-to-one correspondence between the Fermi level and the carrier concentrations. Knowledge of E_F completely specifies n_0 and p_0 and vice versa.

Let us next turn to the nonequilibrium (steady state) situation inside the Problem 1 semiconductor at times $t \gg \tau_p$. For times $t \gg \tau_p$, $\Delta p_n = G_L \tau_p = 10^{11}/cm^3$, $p = p_0 + \Delta p \cong 10^{11}/cm^3$ and $n \cong n_0 = 10^{15}/cm^3$. Although n remains essentially unperturbed, p has increased by many orders of magnitude; and it is clear that the Fig. 3.19(a) diagram no longer describes the state of the system. In fact, the Fermi level is defined only for a system under equilibrium conditions and cannot be used to deduce the carrier concentrations inside a system in a nonequilibrium state.

The convenience of being able to deduce the carrier concentrations by inspection from the energy band diagram is extended to nonequilibrium conditions through the use of *quasi-Fermi levels*. This is accomplished by introducing two energies, F_N, the quasi-Fermi level for electrons, and F_P, the quasi-Fermi level for holes, which are *by definition* related to the nonequilibrium carrier concentrations in the same way E_F is related to the equilibrium carrier concentrations; that is, under nonequilibrium conditions,

$$n \equiv n_i e^{(F_N - E_i)/kT} \quad \text{or} \quad F_N \equiv E_i + kT \ln\left(\frac{n}{n_i}\right) \tag{3.61a}$$

and

$$p \equiv n_i e^{(E_i - F_P)/kT} \quad \text{or} \quad F_P \equiv E_i - kT \ln\left(\frac{p}{n_i}\right) \tag{3.61b}$$

Please note that F_N and F_P are conceptual constructs, with the values of F_N and F_P being totally determined from a prior knowledge of n and p. Moreover, the quasi-Fermi level definitions have been carefully chosen so that when a perturbed system relaxes back toward equilibrium, $F_N \rightarrow E_F$, $F_P \rightarrow E_F$ and Eqs. (3.61) $\rightarrow$ Eqs. (3.60).

To provide a straightforward application of the quasi-Fermi level formalism, let us again consider the $t \gg \tau_p$ state of the semiconductor in Sample Problem No. 1. First, since $n \simeq n_0$, $F_N \simeq E_F$. Next, substitution of $p = 10^{11}/\text{cm}^3$ ($n_i = 10^{10}/\text{cm}^3$ and $kT = 0.026$ eV) into Eq. (3.61b) yields $F_P = E_i - 0.06$ eV. Thus by eliminating E_F from the energy band diagram and drawing lines at the appropriate energies to represent F_N and F_P, one obtains the diagram displayed in Fig. 3.19(b). Figure 3.19(b) clearly conveys to any observer that the system under analysis is in a nonequilibrium state. Figure 3.19(b) referenced to Fig. 3.19(a) further indicates at a glance that low level injection is taking place inside the semiconductor creating a concentration of minority carrier holes in excess of n_i. A second application where the quasi-Fermi levels are a function of position, an application based on Sample Problem No. 2, is left to the reader as an exercise.

As a final point, it should be mentioned that the quasi-Fermi level formalism can be used to recast some of the carrier-action relationships in a more compact form. For example, the standard form of the equation for the total hole current reads

$$\mathbf{J}_P = q\mu_p p \mathscr{E} - qD_P \nabla p \tag{3.62}$$

$$\text{(same as 3.42b)}$$

However, differentiating Eq. (3.61b) gives

$$\nabla p = \left(\frac{n_i}{kT}\right) e^{(E_i - F_P)/kT} (\nabla E_i - \nabla F_P) \tag{3.63a}$$

$$= \left(\frac{qp}{kT}\right)\mathscr{E} - \left(\frac{p}{kT}\right)\nabla F_P \tag{3.63b}$$

Substitution of Eq. (3.63b) into Eq. (3.62) then yields

$$\mathbf{J}_P = q\left(\mu_p - \frac{qD_P}{kT}\right)p\mathscr{E} + \left(\frac{qD_P}{kT}\right)p\nabla F_P \tag{3.64}$$

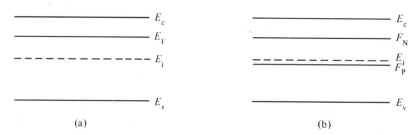

(a)   (b)

Fig. 3.19 Sample use of quasi-Fermi levels. Energy band description of the situation inside the semiconductor of Sample Problem No. 1 under (a) equilibrium conditions and (b) nonequilibrium conditions ($t \gg \tau_p$).

or, making use of the Einstein relationship,

$$\mathbf{J}_P = \mu_p p \; \nabla F_P \tag{3.65a}$$

Similarly,

$$\mathbf{J}_N = \mu_n n \; \nabla F_N \tag{3.65b}$$

Since $\mathbf{J}_P \propto \nabla F_P$ and $\mathbf{J}_N \propto \nabla F_N$ in Eqs. (3.65), one is led to a very interesting general interpretation of energy band diagrams containing quasi-Fermi levels. Namely, a quasi-Fermi level that varies with position ($dF_P/dx \neq 0$ or $dF_N/dx \neq 0$) indicates at a glance that current is flowing inside the semiconductor.

3.6 REVIEW AND CONCLUDING COMMENTS

The examination of sample problems concludes the discussion on carrier action in particular and semiconductor fundamentals in general. Over the course of the volume, the material covered has dealt with a wide variety of items including terms, relevant facts and information, representative numerical values, models or visualization aids, and working equations. With little hesitation the reader should be able to define such terms as unit cell, donor, p-type material, degeneracy, scattering, R – G center, and diffusion length. A more complete collection of fundamental terms is listed in Table 3.1. Useful facts and information were scattered throughout the development. For example, the crystalline structure of Si is described by the diamond lattice, insulators have very large band gaps, and in the major semiconductors $\mu_n > \mu_p$. More extensive information was presented concerning the Miller indexing scheme, the determination of doping type using the hot-point probe, and the measurement of the doping concentration using the four-point probe. A different kind of knowledge, knowledge of representative numerical values, is invaluable in performing crude order-of-magnitude calculations and in gauging the relative size of newly encountered quantities. Table 3.2 lists numerical values for a selected set of key parameters.

   The prime visualization aid described herein was, of course, the energy band diagram. The full description of the diagram actually spanned most of the volume. The basic diagram introduced in Section 2.2 consisted of only two energy levels, E_c and E_v. Donor and acceptor levels were then added to the diagram and were used to help explain the action of dopants in modifying the carrier concentrations. Next the energy levels E_F and E_i were placed on the diagram to indicate at a glance the equilibrium carrier concentrations. Later it was pointed out that the existence of an electric field inside the semiconductor caused band bending or a variation of the energy bands with position. By simply inspecting the diagram one could ascertain the general functional dependence of the electrostatic potential and field present in the material. In the discussion of recombination – generation another level was added near midgap. This level arising from R – G centers plays a dominant role in the thermal communication between the bands. Lastly, quasi-Fermi levels were introduced to describe nonequilibrium conditions.

Table 3.1 List of Terms.

(1) amorphous	(31) extrinsic temperature region
(2) polycrystalline	(32) intrinsic temperature region
(3) crystalline	(33) freeze-out
(4) lattice	(34) drift
(5) unit cell	(35) scattering
(6) ingot	(36) drift velocity
(7) carrier	(37) thermal motion
(8) electron	(38) drift current
(9) hole	(39) current density
(10) conduction band	(40) mobility
(11) valence band	(41) resistivity
(12) band gap	(42) conductivity
(13) effective mass	(43) band bending
(14) intrinsic semiconductor	(44) diffusion
(15) extrinsic semiconductor	(45) diffusion current
(16) dopant	(46) diffusion constant
(17) donor	(47) recombination
(18) acceptor	(48) generation
(19) n-type material	(49) photogeneration
(20) p-type material	(50) direct thermal $R-G$
(21) n^+ (or p^+) material	(51) indirect thermal $R-G$
(22) majority carrier	(52) $R-G$ center
(23) minority carrier	(53) low level injection
(24) density of states	(54) equilibrium
(25) Fermi function	(55) perturbation
(26) Fermi energy (or level)	(56) steady state
(27) nondegenerate semiconductor	(57) quasi steady state
(28) degenerate semiconductor	(58) minority carrier lifetime
(29) charge neutrality	(59) minority carrier diffusion length
(30) ionization of dopant sites	(60) quasi-Fermi level

Working equations developed or derived herein fall into two general categories. One set of equations is used to quantitatively characterize the equilibrium state of the semiconductor. The most important of these equations are cited in the summary at the end of Chapter 2. The second group of equations is used to describe carrier action or model the state of a semiconductor subject to an external perturbation. The latter "equations of state"

Table 3.2 Numerical Values for Key Parameters.

Quantity	Value (Si, T = room temperature)
kT	0.026 eV
E_G	1.12 eV
n_i	$10^{10}/\text{cm}^3$
μ_n and μ_p	1360 and 490 $\text{cm}^2/\text{V-sec}$ (for N_D or $N_A = 10^{14}/\text{cm}^3$)
τ_n and τ_p	$\sim 10^{-6}$ sec. (actual values can range from 10^{-3} to 10^{-9} sec)
L_N and L_P	$\sim 10^{-2} - 10^{-3}$ cm

are reviewed and listed in Subsection 3.4.4. In dealing with either equation set one must always be careful to check the applicability of the equations prior to their utilization.

In conclusion, we have presented and examined those terms, facts, models, equations, etc., that are routinely encountered in semiconductor-related analyses; we have tried to lay a firm foundation for the reader's future adventures into the world of solid state devices.

PROBLEMS

3.1 For temperatures T near room temperature, roughly sketch the expected form of a D_N versus T plot appropriate for *lowly* doped n-type silicon. Record your reasoning leading to the plot.

3.2 (a) A silicon sample maintained at room temperature is uniformly doped with $N_D = 5 \times 10^{15}/\text{cm}^3$ donors. Calculate the resistivity of the sample.

(b) The silicon sample is next "compensated" by adding $N_A = 5 \times 10^{15}/\text{cm}^3$ acceptors. Determine the resistivity of the now compensated sample. (Exercise caution in choosing the mobility values to be employed in this part of the problem.)

3.3 Using the energy band diagram, indicate how one visualizes:

(a) Photogeneration of carriers;

(b) Direct thermal recombination;

(c) Indirect thermal recombination via R–G centers.

3.4 $\partial \Delta n_p / \partial t = D_N \partial^2 \Delta n_p / \partial x^2 - \Delta n_p / \tau_n + G_L$ is known as the minority carrier diffusion equation for electrons.

(a) Why is it called a *diffusion* equation?

(b) Why is it referred to as a *minority carrier* equation?

(c) The equation is valid only under low level injection conditions. Why?

3.5 The energy band diagram for a semiconductor is shown in Fig. P3.5. Sketch the general

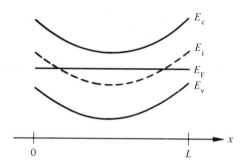

Figure P3.5

form of

(a) n and p versus x,

(b) the electrostatic potential (V) as a function of x,

(c) the electric field as a function of x,

(d) J_N and J_P versus x.

3.6 A semiconductor is characterized by the energy band diagram of Fig. P3.6.

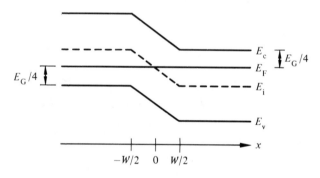

Figure P3.6

(a) Sketch the electrostatic potential (V) inside the semiconductor as a function of x.

(b) Sketch the electric field $(\mathscr{E})$ inside the semiconductor as a function of x.

(c) The semiconductor is in equilibrium. How does one deduce this fact from the energy band diagram?

(d) What is the electron current density (J_N) and hole current density (J_P) at $x = 0$?

(e) Is there an electron *drift* current at $x = 0$? If there is a *drift* current, what is the direction of the current flow?

(f) Is there an electron *diffusion* current at $x = 0$? If there is a *diffusion* current, what is the direction of the current flow?

(g) An electron located at $x = W/2$ attempts to move over to the $x < -W/2$ region of the semiconductor without changing its total energy. What is the minimum kinetic energy (K.E.) the electron must have in order to accomplish this feat?

3.7 The equilibrium and steady state conditions, respectively, before and after illuminating a semiconductor are characterized by the energy band diagrams shown in Fig. P3.7. $T = 300°K$, $n_i = 10^{10}/cm^3$, $\mu_n = 1360$ cm^2/V-sec, and $\mu_p = 490$ cm^2/V-sec. From the information provided, determine:

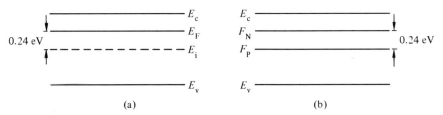

(a)   (b)

Figure P3.7

(a) n_0, the equilibrium electron concentration;

(b) p_0, the equilibrium hole concentration;

(c) p under steady state conditions;

(d) $N_D - N_A$.

(e) Do we have "low level injection" when the bar is illuminated? Explain.

(f) What is the resistivity of the semiconductor before illumination?

(g) What is the resistivity of the semiconductor after illumination?

(h) If the light is generating 10^{16} electrons and holes per cm^3-sec, what is τ_p?

3.8 The $x = 0$ end of a $N_A = 10^{14}/cm^3$ doped semi-infinite bar of silicon maintained at room temperature is attached to a "minority carrier digestor" which makes $n = 0$ at $x = 0$. (*Note:* $n = 0$ and not $\Delta n_p = 0$.)

(a) Characterize the bar under equilibrium conditions by providing numerical values for (i) n_i, (ii) p_0 and (iii) n_0.

(b) What is Δn_p at $x = 0$?

(c) Do we have "low level injection" conditions here?

(d) Determine $\Delta n_p(x)$ in the bar under steady state conditions with the "digestor" attached; i.e., derive a formula for $\Delta n_p(x)$.

3.9 An n-type semiconductor bar of length L (see Fig. P3.9) is maintained under steady state conditions such that $\Delta p_n(0) = \Delta p_{n0} > 0$ and $\Delta p_n(L) = 0$. Here $\Delta p_{n0} \ll n_0$, there are no "other processes" of any type, and *thermal recombination–generation is negligible* at all points inside the bar. Determine $\Delta p_n(x)$ for $0 \leq x \leq L$.

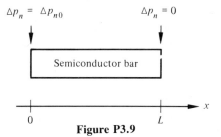

Figure P3.9

3.10 As pictured in Fig. P3.10, an n-type silicon bar of length L contains a linearly graded doping profile varying from $N_D = 10^{15}/cm^3$ at $x = 0$ to $N_D = 10^{17}/cm^3$ at $x = L$; i.e., $N_D(x) = N_D(0) + [N_D(L) - N_D(0)] (x/L)$. Also, the bar is being maintained under equilibrium conditions at room temperature and $N_D(0) \gg n_i$.

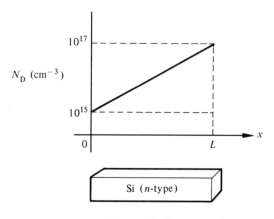

Figure P3.10

(a) Assuming approximate charge neutrality at all points within the bar, derive a formula for $\mathcal{E}(x)$. *Hint:* Review the subsections on charge neutrality and the Einstein relationship (Subsections 2.5.4 and 3.2.4, respectively).

(b) Draw the energy band diagram for the bar. Be sure to show E_F, E_i, E_c and E_v on your E versus x plot.

(c) Although an electrostatic potential drop and an electric field exist within the bar, the current flow in the x-direction at all points must be identically zero. What physical mechanism within the bar offsets the tendency of the carriers to drift in response to the internal electric field?

(d) Roughly (just a general trend desired), how will the mobility of the majority carrier electrons vary as a function of x down the length of the bar?

(e) An acceptor impurity concentration of $9 \times 10^{14}/cm^3$ is now added uniformly throughout the bar's volume. What effect will the added impurity concentration have on the electron concentration within the bar? Explain.

(f) What effect will the added acceptor impurity concentration have on the electron mobility within the bar? Explain.

3.11 Given a semiconductor under steady state conditions, no light, and no "other processes," show quite generally that

$$\mathbf{J} = \mathbf{J}_N + \mathbf{J}_P = \text{constant independent of position (that is, the total current is}$$
$$\text{not a function of } x, y, \text{ or } z).$$

Hint: Start with the continuity equations (Eqs. (3.39)) and make use of the fact that $r_N = r_P$ under steady state conditions.

3.12 Referring to Sample Problem No. 2 of Subsection 3.5.2, nonpenetrating illumination of a semiconductor bar was found to cause a steady state, excess-hole concentration of

$\Delta p_n(x) = \Delta p_{n0} \exp(-x/L_P)$. Under the prevailing low level injection conditions, and since $p = p_0 + \Delta p$, we can then state

$$n \simeq n_0$$

$$p = p_0 + \Delta p_{n0} e^{-x/L_P}$$

for the illuminated sample.

(a) Using Eqs. (3.61), establish relationships for F_N and F_P in the illuminated bar.

(b) Show that F_P is a linear function of x at points where $\Delta p_n(x) \gg p_0$.

(c) Using the results of parts (a) and (b), draw the energy band diagrams describing the semiconductor bar of Sample Problem No. 2 under equilibrium and illuminated steady state conditions. (Assume $\mathscr{E} = 0$ in the illuminated bar.)

(d) Is there a hole current in the illuminated bar under steady state conditions? Explain.

(e) Is there an electron current in the illuminated bar under steady state conditions? (Careful here! Appearances can sometimes be deceiving. In fact, $J = J_N + J_P =$ constant at all points in the bar, according to Problem 3.11, and J clearly vanishes at $x = 0$. Thus, $J_N(x) = -J_P(x)$. The apparent discrepancy stems from the fact that J_N is proportional to both n and ∇F_N. The large size of n requires a very small ∇F_N, and the slope in F_N simply cannot be seen by inspecting the energy band diagram.)

Suggested Readings

General comment: Most texts devoted to a general coverage of solid state devices contain an introductory set of chapters on semiconductor fundamentals. Listed below are four texts which, in the author's opinion, best supplement or enhance the material presented in this volume. These texts contain additional references that might also be consulted by the reader.

R. B. Adler, A. C. Smith, and R. L. Longini, *Introduction to Semiconductor Physics,* Semiconductor Electronics Education Committee (SEEC) Volume 1, New York: Wiley, 1964. SEEC Volume 1 is a work very similar in scope and content to the present volume. It is a bit difficult to read and slightly more advanced in spots, but is generally considered a classic in its field.

D. H. Navon, *Electronic Materials and Devices*. Boston: Houghton Mifflin, 1975. See Chapters 2 to 5. The Navon text contains a more advanced (but still understandable) coverage of the material presented in this volume. It is a text for readers who wish to expand their knowledge of the subject matter.

B. G. Streetman, *Solid State Electronic Devices*. 2nd ed. Englewood Cliffs, N. J.: Prentice-Hall, 1980. See Chapters 1 to 4. The book by Streetman is a good general-purpose text providing a different point of view on many of the topics covered in Volume I.

H. E. Talley, and D. G. Daugherty, *Physical Principles of Semiconductor Devices*. Ames: Iowa State University Press, 1976. See Chapters 1 to 6. This is a fairly readable text that provides a quantum-mechanical and statistical background for some of the concepts and relationships simply introduced herein.

Volume Review Problem Sets and Answers

Volume Review Problem Sets and Answers

To work the following problem sets requires a knowledge — sometimes an integrated knowledge — of the subject matter in all three chapters of Volume I. The sets could serve as a review or as a means of evaluating the reader's mastery of the subject. Assuming "closed-book" conditions and a reader who has thoroughly reviewed the material, Problem Set A should take 20 minutes or less to complete, while Problem Sets B and C were designed to be completed in approximately 50 minutes each. Answers to the problems are included at the end of the Problem Sets.

PROBLEM SET A

A.1 Identify the energy levels labeled E in Fig. PA.1.

(a)   (b)   (c)

Figure PA.1

A.2 Compute the total current flow $(J_N + J_P)$ in the semiconductor characterized by the energy band diagram in Fig. PA.2.

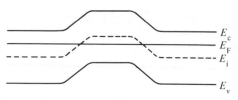

Figure PA.2

A.3 In a silicon crystal, each silicon atom has how many nearest neighbors?

A.4 How much energy ($E_c - E_v$, $E_c - E_D$ or $E_v - E_{ref}$) is required to break a covalent bond in an elemental semiconductor such as silicon?

A.5 Which direction (N, S, E, or W) in Fig. PA.5 corresponds to decreasing electron energy? Which direction corresponds to decreasing hole energy?

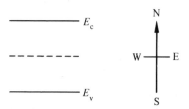

Figure PA.5

A.6 Which semiconductor in Fig. PA.6 is degenerately doped?

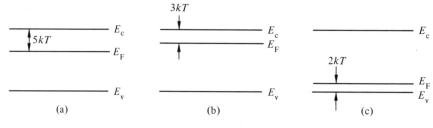

Figure PA.6

A.7 Suppose a semiconductor is doped with an impurity concentration N such that $N \gg n_i$ and all impurities are ionized. Also, $n = N$ and $p = n_i^2/N$. Is the impurity a donor or an acceptor?

A.8 For small electric fields, the average carrier drift velocity is proportional to the electric field. What do we call the constant of proportionality?

A.9 Which carrier (A or B) in Fig. PA.9 has the greatest kinetic energy?

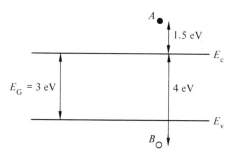

Figure PA.9

A.10 What is the expression for resistivity in an acceptor-doped semiconductor at room temperature?

A.11 We are given $E_v(x)$ in a semiconductor with bent bands. What is the associated electric field?

A.12 Identify the temperature regions A, B, and C in Fig. PA.12.

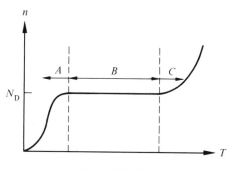

Figure PA.12

A.13 Mobilities in intrinsic material are (higher than, lower than, or the same as) those in heavily doped material.

A.14 One can calculate the energy required to ionize an impurity by modeling the impurity ion and its bound electron (or hole) as a hydrogen-like atom. If E_B is the ionization energy for an impurity and E_H is the ionization energy for hydrogen, then is $|E_B|$ ($<$, $=$, or $>$) $|E_H|$?

A.15 Which Fermi function curve in Fig. PA.15 corresponds to high temperature and which corresponds to low temperature?

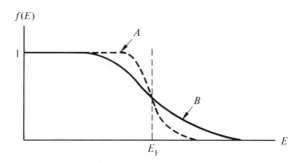

Figure PA.15

A.16 How many valence electrons do the following atoms have? (a) phosphorus; (b) boron; (c) silicon.

A.17 Which semiconductor shown in Fig. PA.17 will have the greatest *intrinsic* carrier concentration at room temperature?

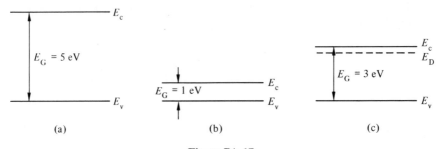

Figure PA.17

A.18 Suppose we know that the electron mobility in Si is 1350 cm²/V-sec. at room temperature. What is the electron diffusion constant?

A.19 Light is used to create excess carriers in silicon. These excess carriers will predominantly recombine via (direct, indirect, or photo) recombination.

A.20 What is the algebraic statement of low level injection?

PROBLEM SET B

B.1 Answer the following questions as concisely as possible.

(a) A material is said to be amorphous. What does this mean?

(b) A certain material crystallizes in a lattice structure characterized by a simple cubic unit cell. The side length of the unit cell is $a = 5 \times 10^{-8}$ cm. How many atoms/cm³ are there in the material?

(c) Given that E_F is positioned at E_c, determine (numerical answer required) the probability of finding electrons in states at $E_c + kT$.

(d) Using the measured resistivity, you plan to deduce a sample's doping concentration. Why is it necessary to also know the sample's doping type (n or p) before determining the doping concentration?

(e) Using the energy band diagram, indicate how you would visualize:

(i) Freeze-out of majority carrier electrons at donor sites as the temperature is lowered toward 0°K.

(ii) Capture of a conduction band electron at a R-G site.

B.2 A semiconductor is characterized by the energy band diagram shown in Fig. PB.2. It is also known that $E_G = 1.12$ eV, $kT = 0.026$ eV, $n_i = 10^{10}/\text{cm}^3$, $\mu_p = 470$ cm^2/V-sec and $\tau_p = 10^{-6}$sec.

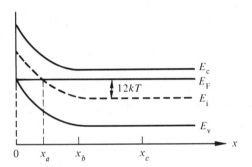

Figure PB.2

(a) Sketch the electrostatic potential (V) inside the semiconductor as a function of x.

(b) Sketch the electric field ($\mathscr{E}$) inside the semiconductor as a function of x.

(c) Is the semiconductor degenerate at any point? If so, where?

(d) Sketch the hole concentration as a function of position, specifically indicating on your plot numerical values for p at $x = x_a$ and $x = x_c$.

(e) (i) Is there a hole drift current at $x = x_a$?

(ii) Is there a hole diffusion current at $x = x_a$?

(iii) What is the total hole current density (J_P) at $x = x_a$?

(f) A small excess of holes is introduced at the point x_c. If $x_c - x_b = 10^{-2}$cm, will a significant number of the excess ever reach x_b? Explain.

B.3 A semi-infinite p-type bar (see Fig. PB.3) is illuminated with light which generates G_L electron–hole pairs/cm^3-sec uniformly throughout the volume of the semiconductor. Simultaneously, carriers are extracted at $x = 0$ making $\Delta n_p = 0$ at $x = 0$. Assuming that a steady state condition has been established and $\Delta n_p(x) \ll p_0$ for all x, determine $\Delta n_p(x)$.

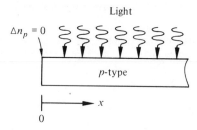

Figure PB.3

PROBLEM SET C

C.1 Answer the following questions as concisely as possible.

(a) How many atoms are there in a body centered cubic unit cell?

(b) Draw a picture showing how one visualizes a donor in the bonding model for a semiconductor.

(c) An average hole drift velocity of 10^3cm/sec results when 2V is applied across a 1-cm-long semiconductor bar. What is the hole mobility inside the bar?

(d) Two bars of the same material, one bar n-type and one p-type, are homogeneously doped such that N_D(bar 1) $= N_A$(bar 2) $\gg n_i$. Which bar will exhibit the larger resistivity?

(e) Using the energy band model, indicate how one visualizes:

   (i) Direct thermal generation.

   (ii) Indirect thermal generation via R–G centers.

C.2 A semiconductor under equilibrium conditions is characterized by the energy band diagram of Fig. PC.2. Given $E_G = 1.12$ eV, $n_i = 10^{10}$/cm³ and $kT = 0.026$ eV

(a) Determine n and p at $x = L/2$.

(b) Determine n at $x = L/4$.

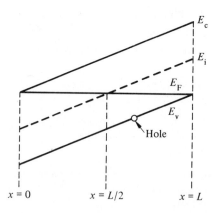

Figure PC.2

(c) For what values of x, if any, is the semiconductor degenerate?

(d) What is the approximate kinetic energy of the hole shown in the diagram?

(e) Sketch the electric field ($\mathcal{E}$) inside the semiconductor as a function of x.

(f) If $L = 1$ cm, what is the magnitude of the electric field inside the semiconductor?

C.3 (a) Listed below are modified forms of the electron minority carrier diffusion equation. Indicate what special conditions must exist in the physical system under analysis to obtain the specific form quoted.

   (i) $d^2 \Delta n_p/dx^2 = 0$

   (ii) $0 = D_N \, d^2 \Delta n_p/dx^2 - \Delta n_p/\tau_n + G_L$

   (iii) $d\Delta n_p/dt = -\Delta n_p/\tau_n$

(b) A piece of silicon ($N_A = 10^{15}/cm^3$, $\tau_n = 1\mu sec$) maintained at room temperature is first illuminated for a time $t \gg \tau_n$ with light which generates $G_{L0} = 10^{16}$ electron-hole pairs per cm^3-sec uniformly throughout the volume of the silicon. At time $t = 0$ the light intensity is reduced, making $G_L = G_{L0}/2$ for $t \geq 0$.

   (i) Do we have low level injection here? Explain.

   (ii) Determine $\Delta n_p(t)$ for $t \geq 0$.

ANSWERS—SET A

A.1   (a) Trap (R–G center) level; (b) donor level; (c) acceptor level.

A.2 $J = 0$ (E_F is position independent. — Thus equilibrium conditions prevail.)

A.3   4

A.4 $E_c - E_v$

A.5 S for decreasing electron energy; N for decreasing hole energy.

A.6   semiconductor (c)

A.7   donor

A.8   mobility

A.9   carrier A

A.10 $\rho = 1/q\mu_p N_A$

A.11 $\mathcal{E}(x) = (1/q)(dE_v/dx)$

A.12 A = freeze out; B = extrinsic T-region; C = intrinsic T-region.

A.13 higher than

A.14 $|E_B| < |E_H|$

A.15 A = lower T, B = higher T

A.16 (a) phosphorus—5; (b) boron—3; (c) silicon—4.

A.17 semiconductor (b)

A.18 $D_N = 35.1 \text{ cm}^2/\text{sec}$

A.19 indirect recombination

A.20 $n \simeq n_0$, $\Delta p \ll n_0$ in n-type material; $p \simeq p_0$, $\Delta n \ll p_0$ in p-type material.

ANSWERS — SET B

B.1 (a) The atomic arrangement in the material exhibits no long-range order.

(b) 8×10^{21} atoms/cm³

(c) $f(E_c + kT) = 0.269$

(d) The doping concentration is determined from a ρ versus N_A or N_D plot and there are two different plots, one for n-type and one for p-type material. (ρ depends on the carrier mobility which is different for n- and p-type materials with the same doping.) Thus one must know the type to choose the proper ρ versus doping plot.

(e)

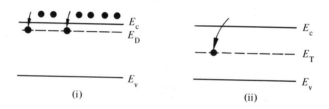

(i)   (ii)

B.2 (a)

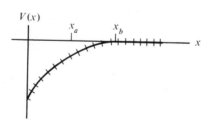

(b)

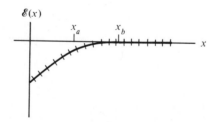

(c) The semiconductor is degenerate at and near $x = 0$.

(d)

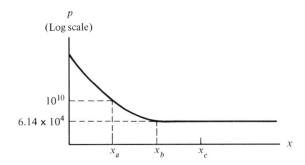

(e) (i) Yes... there is a drift current because $p \neq 0$ and $\mathcal{E} \neq 0$.

(ii) Yes... there is a diffusion current because $dp/dx \neq 0$ as seen in part (d)

(iii) $J_P = 0$... under equilibrium conditions the drift and diffusion components cancel out

(f) $L_P = 3.50 \times 10^{-3}$cm and $(x_c - x_b)/L_P = 2.86$. Since the point x_b is some 2.86 diffusion lengths from x_c, relatively few of the excess holes will make it to x_b.

B.3 ... $\Delta n_p(x) = G_L \tau_n (1 - e^{-x/L_N})$

ANSWERS—SET C

C.1 (a) two atoms

(b) Draw Fig. 2.10(a)

(c) 500 cm^2/V-sec

(d) $p(\text{bar } 2) > p(\text{bar } 1)$

(e) (i) Draw the right-hand side of Fig. 3.14(b).

(ii) Draw the right-hand side of Fig. 3.15(b).

C.2 (a) $n = p = n_i = 10^{10}$/cm^3

(b) $n = 4.75 \times 10^{14}$/cm^3

(c) The semiconductor is degenerate for $0 \leq x \leq \Delta L$ and $L - \Delta L \leq x \leq L$, where $\Delta L = 3kTL/E_G$.

(d) K.E. ≈ 0

(e) $\mathcal{E} = $ constant for all x

(f) $\mathcal{E} = (1/q)(E_G/L) = 1.12$ V/cm

C.3 (a) (i) Steady state, no thermal R–G, no photogeneration

(ii) Steady state

(iii) No concentration gradient, no photogeneration

(b) (i) We do have low level injection.

$$\Delta n_{p|max} = G_{L0}\tau_n = 10^{10}/\text{cm}^3 \ll p_0 = N_A = 10^{15}/\text{cm}^3$$

(ii)
$$\Delta n_p(t) = \frac{G_{L0}\tau_n}{2}(1 + e^{-t/\tau_n})$$

Appendix

Appendix

LIST OF SYMBOLS

A	area
c_n	electron capture coefficient
c_p	hole capture coefficient
D_N	electron diffusion constant (cm^2/sec)
D_P	hole diffusion constant (cm^2/sec)
$\mathscr{E}$, $\mathscr{E}$	electric field
E	energy
E_A	acceptor energy level
E_B	binding energy at dopant (donor, acceptor) sites
E_c	lowest possible conduction band energy
E_D	donor energy level
E_F	Fermi energy or Fermi level
E_G	band gap or forbidden gap energy
E_H	electron binding energy within the hydrogen atom
E_i	intrinsic Fermi level
E_T	trap or R–G center energy level
E_v	highest possible valence band energy
$\mathbf{F}$	force
$\mathscr{F}$	four-point probe correction factor
$f(E)$	Fermi function
F_N	quasi-Fermi level (or energy) for electrons

F_P	quasi-Fermi level (or energy) for holes
$F_{1/2}(\eta)$	Fermi integral of order 1/2
$g_c(E)$	density of conduction band states
G_L	photogeneration rate; number of electron–hole pairs created per cm^3-sec
g_N	net general generation rate of electrons due to "other processes"
g_P	net general generation rate of holes due to "other processes"
$g_v(E)$	density of valence band states
h	Planck's constant
$\hbar$	$h/2\pi$
I	current
I_P	hole current
$I_{P|drift}$	hole current due to drift
$\mathbf{J}, J$	current density (amps/cm^2)
$\mathbf{J}_N, J_N$	electron current density
$\mathbf{J}_{N|diff}$	electron current density due to diffusion
$\mathbf{J}_{N|drift}$	electron current density due to drift
J_{Nx}, J_{Ny}, J_{Nz}	x, y and z direction components of the electron current density
$\mathbf{J}_P, J_P$	hole current density
$\mathbf{J}_{P|diff}$	hole current density due to diffusion
$\mathbf{J}_{P|drift}$	hole current density due to drift
J_{Px}, J_{Py}, J_{Pz}	x, y and z direction components of the hole current density
k	Boltzmann's constant (8.62×10^{-5} eV/°K)
K.E.	kinetic energy
K_S	semiconductor (usually Si) dielectric constant
L_N	electron minority carrier diffusion length
L_P	hole minority carrier diffusion length
m_n^*	electron effective mass
m_0	free electron mass
m_p^*	hole effective mass
n	electron carrier concentration (number of electrons/cm^3)
n^+	heavily doped n-type material
N_A	total number of acceptor atoms or sites/cm^3
N_A^-	number of ionized (negatively charged) acceptor sites/cm^3
N_C	effective density of conduction band states
N_D	total number of donor atoms or sites/cm^3
N_D^+	number of ionized (positively charged) donor sites/cm^3
n_i	intrinsic carrier concentration

n_0	equilibrium electron concentration
N_T	number of R–G centers/cm^3
N_V	effective density of valence band states
p	hole concentration (number of holes/cm^3)
p^+	heavily doped p-type material
P.E.	potential energy
p_0	equilibrium hole concentration
q	magnitude of the electronic charge (1.6×10^{-19} coul)
r_N	net general thermal recombination rate for electrons
r_P	net general thermal recombination rate for holes
s	probe-to-probe spacing in a four-point probe
t	time
T	temperature
$\mathbf{v}$	velocity
V	voltage; electrostatic potential
$\mathbf{v}_d, v_d$	drift velocity
Δn	$\Delta n = n - n_0$; deviation in the electron concentration from its equilibrium value
Δn_p	Δn in p-type material
Δp	$\Delta p = p - p_0$; deviation in the hole concentration from its equilibrium value
Δp_n	Δp in n-type material
ε_0	permittivity of free space (8.85×10^{-14} farads/cm)
η	$\eta = (E - E_c)/kT$
η_f	$\eta_f = (E_F - E_c)/kT$
η_f'	$\eta_f' = (E_v - E_F)/kT$
μ_n	electron mobility
μ_p	hole mobility
ρ	resistivity (ohm-cm) or charge density (coul/cm^3)
σ	conductivity
τ_n	electron minority carrier lifetime
τ_p	hole minority carrier lifetime

Index

Index